ENVIRONMENTAL
by DESIGN

ENVIRONMENTAL *by* DESIGN

VOLUME I: INTERIORS

A Sourcebook of Environmentally Aware Choices

KIM LECLAIR &
DAVID ROUSSEAU

H&M
PUBLISHERS

Published simultaneously in the U.S.A. and Canada by
Hartley & Marks, Inc.
79 Tyee Drive, Point Roberts WA 98281

and

Hartley & Marks, Ltd.
3661 West Broadway, Vancouver, BC V6R 2B8

Printed in the U.S.A. on recycled paper.

ISBN 0-88179-085-0

If not available at your local bookstore,
this book may be ordered from the publisher.
Send the cover price plus two dollars for
shipping to either of the above addresses.

LIBRARY OF CONGRESS CATALOGING-IN-PUBLICATION DATA

Leclair, Kim, 1961–
Environmental by design: a sourcebook of environmentally
conscious choices for homeowners, builders & designers / by Kim
Leclair & David Rousseau.
p.   cm.
Includes bibliographical references and index.
ISBN 0-88179-085-0 (pbk.)
1. Environmental protection—Citizen participation.
2. Environmental protection—Equipment and supplies—Catalogs.
3. Building materials—Catalogs.   4. Household supplies—Catalogs.
5. Consumer education.   I. Rousseau, David.   II. Title.
TD171.7.L43   1992
696—dc20   92-37314
                                                CIP

NOTE

All product information in this guide was collected from producer sources and from
research data available in the public domain. It is not intended to be an exhaustive survey of
all materials or sources, nor is any product warranty expressed or implied by listing.
The authors, the publishers, and their agents accept no responsibility for
omissions or for inaccurate data provided by those surveyed.

The health toxicity discussions in this book do not constitute advice on the part
of the authors, publishers, or their agents. The appropriate use of
products and their compatibility with other materials must
be determined by the reader.

CONTENTS

ACKNOWLEDGMENTS

For their ideas and assistance with the making of this book, special thanks are
due to:

Russel Black
Lauren Cavallin
Jessica Chadwick
Sandy Fraser
Stephen Gregory
Stephen Hardy
Keith Jakobsen
Moreka Jolar
Elizabeth McLean
Nicole Milkovich
Darren Onyskiw
Joe Wokolsky

The authors wish to gratefully acknowledge the Foundation for Interior
Design Education Research for its contribution to this guide's research costs
through a Joel Polsky research grant.

FOREWORD

This guide was inspired by the frustration of knowing that there is much more to consider when choosing interior finishing materials than cost and color, but no comprehensive resource for making better health and environmental choices. Even though there is a wealth of available information, it is widely scattered and difficult to work with or apply on a daily basis. Monthly newsletters, independent research articles, and manufacturers' and industry claims abound, but we felt that an informative guide was needed to help us make fair comparisons and educated choices.

As design professionals and consumers, our choices of materials entail many more environmental and health repercussions than we might realize. These are significant on even a small scale, but they become undeniably important when considering large office and retail buildings, and multiple housing developments. With our growing concern for a healthy environment and the environmental challenge to our own health, designers, consumers, and building product manufacturers have an essential role to play. And companies with safe products and environmentally conscious practices should be recognized and supported for taking action.

In our search for information, we began by approaching manufacturers who had established an environmentally concerned reputation or who were promoting their products as having some environmental or health benefit. In the process we discovered many more, and also reviewed those with which we have had long experience. The resulting products listed in this sourcebook range from traditional materials to those at the leading edge of technology.

We surveyed each product with a questionnaire which addresses issues from the cradle to the grave of a product's life cycle. Important aspects of each stage of production and use were identified and represented with a visual symbol. These symbols were then defined, and a minimum compliance threshold established. While some are common considerations in the choice of materials (durability, maintenance requirements), many others have seldom been considered by designers and consumers. Still others deal with contemporary, controversial issues which have legal and ethical implications.

Much research has been done about building materials on a generic level, relating to product performance and manufacturing standards, but little information is available for individual products with environmental and health consequences and company practices taken into account. Our primary purpose is to raise general awareness of these concerns in relation to materials choices, and to introduce some of the alternatives.

In this book we have included two case scenarios of the environmental and health impacts of typical small renovations; the Product Report pages, each section of which includes a brief discussion of the issues for those industries (e.g. flooring, furniture), and a list of available materials with some special environmental merit; together with a glossary and a suggested list for further reading.

We believe that this book will begin to bridge the gap between environmentally conscious personal actions, and individual and professional actions within the building industry. When we choose better products and practices, our built environment is enriched, and we and our natural environment are protected.

All acts have an "environmental cost," and nearly everything is toxic to some degree. We can find or do almost nothing which is perfectly benign for our earth and our health. The important measure of any choice is the degree to which the environmental and health costs have been minimized, and the real benefits realized. Better choices of building materials mean having less impact on the earth while providing pleasing and healthy places to live and work.

Until the consumer understands important issues like those covered in this sourcebook, and is willing to pay the "real cost" of quality, industry will only make changes when forced to do so by regulation.

We hope that this book will provide some new insights for you and help you to make more environmentally sound choices at home and at work.

In the spirit of care and conservation which is becoming more necessary for us all,

Kim Leclair
David Rousseau
1992

For those manufacturers or product representatives who would like a product considered for listing in future editions, please find a reply form at the back of the book.

For those design or construction professionals who would like a subscription to a looseleaf version of this book, including periodic updates, professional source listings, and a newsletter service, please find a subscription form at the back of the book. This bound edition can also be updated to the looseleaf, professional version by returning it to the authors for a partial rebate on the subscription price.

INTRODUCTION

• ENVIRONMENTAL, HEALTH & ETHICAL ISSUES •

This book is a guide to available interior materials and finishes that have some special environmental and health merit. Their production and use mean less energy consumption, less natural resource depletion and pollution, and they are generally less toxic than many conventional materials. Perhaps even more importantly, this book defines some of the issues surrounding the use of materials, which we must face before we can ensure a more sustainable and healthy future.

THE ISSUES

The issues around the building products described in this book fall into two main categories: the condition of our global environment and our own health. Diverse and detailed research has been done in both categories, but little of it is accessible to consumers. The following pages contain a general overview of both issues, as well as a third which deals with the social and ethical aspects of doing business.

THE GLOBAL ENVIRONMENT

We need to recognize the connections between our actions (and inaction) in the building industry and the state of the world. Because access to this information has been poor, many decisions are made without knowledge of, or regard for the global picture.

The building products which we have been using and, indeed, the built environment itself, are directly and indirectly linked with many serious types of global impacts: global warming, acid rain, emissions causing depletion of the ozone layer, toxic waste sites, and growing landfills. These impacts result from the extraction of raw materials, their processing, shipping, installation, use, maintenance, and, eventually, disposal.

That we can foresee the limits of many nonrenewable energy and material resources, and the destruction of once-thriving ecosystems is now being acknowledged. We know that these limits are, in part, due to the exploitation required to satisfy demands for consumer and building products, exotic and traditional—demands which have grown without the awareness required for sustainability. This demand-driven situation, and the use of high volume, highly mechanized manufacturing processes with their high energy consumption and low labor requirements affect the earth and all its people economically, socially, and environmentally.

Environmental depletion often translates into buyer's language as increased prices, decreased availability, and inferior quality of natural materials. Most often these are seen as inconveniences and not as the result of an overall problem which can be altered by use of the consumer's buying power.

Energy intensity (the total amount of energy use represented by the manufacture of a product) and other environmental costs of a product are price factors that have not traditionally been passed along to the end-user. While some consider recycled paper, for example, as "more expensive," the full cost of "regular" paper is not included in the purchase price. If reduced forests for pulp supply, contaminated river systems, air pollution, and a variety of illnesses were included, the price would look very different. Though environmental regulations restrict emissions, they are usually inadequate or ill-enforced due to the pressure to maintain production.

In these pages some of the hidden environmental costs have been included in our evaluations.

INTERIOR CONSTRUCTION MATERIALS AND OUR HEALTH

There are several ways in which the chemical composition of a material affects our health: first, there is the exposure to people who produce it; second, there are the health effects of handling and installation; third, there are the effects of maintenance; and finally, of removal and disposal.

Nearly everyone is now also aware that the environment inside buildings may not be beneficial to our health and well-being. Poor air quality, poor lighting, poor furniture design, and a number of other problems contribute to sick building syndrome, building-related illness, and unhealthy interiors. All these terms are now familiar. The effects of exposure to toxic materials and compounds in indoor air have been widely observed, but little has been said, until recently, about the contribution of the choice of interior materials. Most office buildings have ventilation systems designed to exhaust emissions from interior materials and other sources, and to bring in outside air, but some do not function as designed or are overloaded. Most homes do not have ventilation systems.

In any case, technology cannot be relied upon alone for a solution. As in preventative medicine, dealing with problems (emissions) at the source can significantly improve indoor air quality and reduce resulting complaints and possibly illness. Reducing pollution at the source will work both with or without ventilation, and can help make ventilation more effective.

Many air contaminants are produced by adhesives, finishing products, and cleaning compounds used during installation, exposing trades and maintenance people to unhealthy high concentrations. Other contaminants are produced as these cure, and as materials offgas, putting the occupants at risk. And, while some materials stabilize quickly, others, though their emissions may be low, take much longer to diminish.

The major pollutants from indoor materials are volatile solvents, formaldehyde, and other long-term volatile emissions, biological particles (molds and bacteria), and other particles (mineral fibers, dusts). These are found, to some degree, in most everyday materials, either due to their composition or their condition as they become contaminated or wear out. Their long-term effects can be varied, and include such complaints and illnesses as chronic respiratory and eye inflammation, asthma, skin rashes, fatigue, depression, excessive colds and flu. These are conditions which seem to afflict us more and more frequently, probably for a number of reasons, and exposure to poor indoor air is one we can do something about.

In these pages some of the health implications of production, installation, and use have been included in our evaluations.

SOCIAL AND ETHICAL ISSUES

Beyond the environmental and health effects of materials, there are associated social and ethical issues. For example, some industries provide very little fair and secure employment to their employees, while others do. There are many active steps which a manufacturer can take to improve social conditions and to reinvest profits for the benefit of all. For example, a company can provide good benefit packages and profit sharing plans for employees. It can invest substantially in research and development, and education.

Some companies are involved with weapons production, or the use of nuclear power, while others are not. Considering all the unknowns about the future of spent nuclear fuel and obsolete, contaminated plants, nuclear power is a good example of a technology that appears to be affordable today because the real costs and ecological impact are being deferred to tomorrow.

Although most manufacturers of building products are not directly involved in toxicity testing on animals, the suppliers and producers of their raw materials often are. Some companies do very little or no toxicity testing, either by policy or because their products are regarded as safe through traditional use. In some cases, toxicity testing is mandatory to having a product approved by government agencies.

Only toxicity and combustion hazard testing uses animals. Durability, strength, colorfastness, stain resistance, and acoustics tests do not require the use of animals. Mice, rats, rabbits, and dogs are exposed to toxic or combustion gases produced by a material in order to establish the effect of a measured dose. While not all animals die during actual testing, they do suffer, and are destroyed at the end of experimentation.

Alternative testing methods, which do not use animals and meet government standards, have been developed in the cosmetics industry. However, these methods have been used very little in other industries. Some manufacturers will make the results of toxicity tests available upon request. Asking for these results is the surest method of determining whether animal testing has been performed.

In these pages some of the producer companies' social and ethical concerns have been included in our evaluations.

INFORMED CHOICE

The heart of this book is in the questions it raises and the methods proposed for making informed choices. There are really no right answers or best choices—only a better informed way of making decisions. Consumers ultimately have to make choices based on their own best evaluation of the information available.

IS NATURAL BETTER?

There is a good deal of emphasis on natural materials in this book, and the use of natural materials would seem to be an obvious esthetic choice in an age of increasing environmental awareness. If natural materials are defined as "those taken from nature with the least intervention of manufacturing processes and synthetic chemistry," then there are some sound environmental and health arguments in their favor.

- Many natural materials, such as wood, and plant oils, are potentially renewable. Others, such as sand, though not renewable, occur in vast quantities in nature.
- Generally the less manufactured a material is, the less air and water pollution, and energy use are associated with it.
- Many natural materials have lower toxicity than their synthetic counterparts. Some have been proven safe over centuries, while new synthetics may or may not prove safe over time.

The use of natural materials is not a panacea, however. All materials exact some environmental toll in their production, and some natural materials such as lead and asbestos are well-known health risks. On balance, though, it can be said that natural materials (those which are not highly manufactured or synthesized) from renewable sources are more sustainable, less damaging to the natural environment, and often better for health than synthetic materials.

A NOTE ON RECYCLING

Many people know of the three Rs: Reduce, Reuse, and Recycle. Some have suggested adding others such as Rethink. It is no accident that the first of the three Rs is Reduce. If demand for material goods continues to increase at today's rate, limits will be reached very quickly. All that our best recycling measures can do is to slow the onset of those limits.

In theory, recycling is very good for some metals like aluminum, moderately good for paper, and poor for some materials like glass and plastics. In practice, most recycling is another story. Probably the only real recycling success story is for pure metals which are not mixed with other materials. Recent efforts to collect recyclable paper, plastic and glass, for example, are hampered by inadequate markets, a lack of processing facilities, and high storage and transportation costs. The market value of recyclable paper in many regions is now so low that it is difficult to find anyone to take it.

There are really two types of problems here. One is that industries like paper production are so heavily committed to new materials that they are very slow to shift to recycled products. Another is that the energy and pollution benefits of some types of recycling, such as glass, are marginal at best. They are being done at a net environmental loss just to keep material out of the solid waste stream.

A different type of problem is posed by the chemical complexity of materials like plastics which must be kept separate for recycling. However, most are mixed with other materials during manufacturing, making them difficult to separate. The result of this problem is that what little plastic recycling is now done produces a very low grade product with limited usefulness.

We consider the use of recycled contents and efficiency of manufacture as having the greatest effect on reducing environmental impact, and we see recyclability as a potential which may or may not be fully realized.

POSITIVE ACTION

DESIGNERS

As primary consultants in the development of both new buildings and renovation/rehabilitation projects, a designer is responsible to both the client and the environment. While informed choice on the basis of environmental impact is only now beginning to appear in practice, ignorance in the area of the health impacts of materials is already a very real liability. Fostering conservation, and creating healthy, safe, working and living environments is a professional's mandate.

Some of the actions open to designers are:
• Educating those involved (clients, trades, suppliers, and other consultants) by proposing and selecting safer alternatives.
• Asking sales representatives for information about the safety and environmental characteristics of their products. Letting suppliers know that these are important matters.
• Supporting manufacturers and trades with environmentally conscious business practices, upgraded facilities, and environmentally sound products. Supporting those who are taking the necessary and sometimes costly steps toward change.
• Knowing material properties and using them optimally, to reduce replacement and waste.
• Renovating buildings, as well as salvaging and reusing building products and materials.
• Specifying the use of recycled materials and materials which are currently being recycled.
• Weighing the *actual* cost of materials (including the health and environmental costs) rather than just the purchase cost.
• Weighing the relative merits of materials, with the most appropriate material for a specific use. Selecting materials for their longevity, durability, and ease of maintenance.
• Supporting local industry to avoid excessive shipping, to benefit local economies, and to encourage local environmental action.

INDIVIDUAL CONSUMERS

Whether acting independently or working with a building professional, as the end-users consumers can be the most powerful force for introducing environmental, health, and social criteria in the selection of materials. Ignorance on the subject can be very costly in terms of health and productivity, as well as for the natural environment.

Knowing the issues surrounding indoor air quality and environmental impacts will help to achieve healthier interiors, and provide protection from less aware designers and contractors.

Some of the actions open to consumers are:

- Asking questions about where materials and products came from, and what the contents are. Adding these to the list of criteria when considering a purchase.
- Identifying allergic responses to materials for yourself and members of your family and employees.
- Expressing your concerns to your architect, interior designer, or contractor, and requesting that they consider environmental and health criteria when choosing materials.

ABOUT THE PRODUCT REPORTS

This sourcebook is intended to provide information relating to a wide range of building industries, and to serve as a practical working tool. It is aimed at consumers, designers, building-related consultants and trades. All materials included in these pages are interior construction and finishing materials. They do not include building structural materials or building systems, but begin with insulations and interior sheet materials, and follow through to floor coverings, paints, and furnishings.

The product reports describe products and manufacturers dealing with a variety of important environmental and health concerns. In order to produce a practical guide, the product reports include a wide range of information about each product's entire life cycle, and about the manufacturing company's policies and practices.

The manufacturers that we approached for inclusion in this book generally fall into three categories: those who continue to produce traditional, natural products; those who are effectively and efficiently using a byproduct or recycled materials as their primary resource; and those who are promoting their products as having some environmental or health benefit through technological advances. The responses and information obtained from manufacturers in each of these categories were quite different from one another.

NATURAL MATERIALS

The manufacturer's environmental statement, for example, becomes less important for those using traditional materials, which are usually obtained and produced in a sustainable manner. Since the manufacturers had not abandoned these products in the face of advancing synthetic technology, the lessened environmental impact and other merits have become evident. These materials have once again come into their own and, indeed, when one natural paint distributor was asked about his company's environmental statement, he countered that it was not they who needed an environmental statement, but the multinational corporations who manufacture products that pollute and deplete resources.

Some of these natural products are marketed to people with chemical sensitivities and environmental illnesses, and trading often occurs at a grassroots level. Information was easy to obtain, as manufacturers and distributors are eager to educate, and to share their knowledge about their products.

The simple but elegant statement that the product was the result of the need to reduce waste byproducts of industry and post-consumer waste is both valid and admirable in a society driven by disposability, and with a continual quest for the new. While only one visual symbol is allotted to "recycled content" on the product reports, the actual impact of that criterion alone is substantial.

We found that manufacturers of products in this category were matter of fact, not usually with an environmental mission, so much as having the good sense to recognize that the waste of one industry can be a more than suitable resource for another industry. The implementation of such practices and the development of a market for the product indicate the financial benefits of both resource efficiency and a conservationist attitude.

HIGH TECHNOLOGY

For industries where production had advanced to the point of having high environmental and health impacts, further research was then required to find solutions to problems that technology itself had created. In the meantime, some of these products had become so integral to our building methods that the advanced solutions, although essential, raise the question of the validity of the standard use of the material in the first place.

It was more difficult to extract information from these manufacturers, since in the realm of patents and proprietary ingredients secrecy is their market advantage. The intricacies of replacement compounds, always claimed to be more benign, or low toxicity, must be evaluated in context for their actual effect on plant emissions, or the actual reduction in health risk. This evaluation must be done not only in comparison to the more toxic material, but also to a more traditional, nontoxic material that has been proven through generations of use. The high technology materials, though, usually have a benefit in the aspects of durability and low maintenance. These criteria are very important in reducing the replacement of materials, and thus slowing down landfill accumulation.

CRITERIA

The field of environmental and health criteria for evaluating products and materials is still in its infancy. The criteria used in this guide reflect the current state of understanding of the main issues.

The questionnaire, our basic tool for collecting manufacturers' information, deals with each product's environmental and health impact at all stages of its life cycle, as well as the company's policies and practices. Material Safety Data Sheets, provided by manufacturers as a requirement under occupational health and safety regulations, other product literature from the manufacturers, as well as independent research on materials were used in evaluating the products.

Where information was not available, we assumed the criterion was not met and no credit was given. We have not included products which did not have merit in at least two categories, or for which claims were unsubstantiated or were in conflict with public information.

The visual symbols on the product report pages are explained below, together with the criteria that were met when a symbol was highlighted.

Though no "negative marks" are given, the standards used are very high, and even the lowest impact products receive credit in only seven or eight of the fourteen criteria areas. Caution should be used in comparing product reports, as this is not an absolute scoring system. For example, some of the criteria may not apply to a particular industry. The purpose of the reports is to provide information on, and highlight, the merit of each product.

THE VISUAL SYMBOLS USED IN THIS BOOK

PRODUCTION

Much of the environmental and health impact occurs during the raw materials and manufacturing stages of a product before it ever reaches its final destination. These are also the stages about which designers and consumers know the least. Listed below, with their matching symbols, are the most important quantifiable aspects of each type of product's manufacture. They include every step, from the obtaining of raw materials to primary and secondary manufacturing.

Products with some favorable qualities at these stages are recognized under the following categories:

The adverse environmental impact of first-time production of a material is substantially lessened by using recycled content. Recycled contents reduce the consumption of nonrenewable energy and the limited reserves of raw materials. They also reduce the quantity of solid waste accumulating in landfills.

In this book, "recycled content" refers to the use of post-consumer waste or industrial scrap material from outside the production process, as raw material for manufacturing. The word "recycled" is currently used for everything from the use of in-plant mill scraps to the reuse of salvaged building components. The use of in-plant scrap is not considered recycling under our definition but can be credited as process efficiency. (See RESOURCE RECOVERY, p. 12, for further discussion.)

Definition of our criteria: Credit is given to those products for which post-consumer waste content or use of industrial waste from outside the production process is technically feasible and either already exceeds 10%, or where the manufacturer has made significant gains in the past five years. In-plant reuse of process waste does not qualify.

We now see an end to many of the planet's easily accessible raw materials. There is an urgent need to both lessen our consumption, and to renew these materials so as to ensure their continuing availability. Only renewable resources and those which have reserves so vast as to be practically inexhaustible, such as sand, can be sustainably managed. If we continue to waste and exploit nonrenewable resources like oil and timber, the remaining reserves will lead first to even more limited and expensive supplies, and to their eventual disappearance. It is very difficult to estimate how much remains of most raw materials, to predict what substitutes may be found, or to measure for how long they will still be available. Because of this we have listed manufacturers who have chosen their source from the best available managed resource, or from nonrenewables which are plentiful.

Definition of our criteria: We give credit to products with renewable contents, and which come from sources with strict harvesting controls and regeneration programs. For products with nonrenewable contents, extraction management and mine site restoration practices must either be state of the art, or have improved significantly in the past five years.

Energy intensity is the amount of energy required to produce an item. As most of the energy used comes from burning fossil fuels or flooding land for dams and hydroelectric plants, it is another indicator of a product's environmental impact. With greater manufacturing energy efficiency (less energy required, or the recycling of waste heat), this impact can be lessened. Recognition is given for in-plant efficiency measures such as motor and control improvements, heat recovery, cogeneration (where both steam and electricity may be produced by the plant boiler), certain waste-to-fuel measures, and other similar steps to reduce energy input to products.

Definition of our criteria: We give credit to products for which in-plant energy efficiency is either state of the art or where significant improvements have been made in the past three years.

In Europe, plant emissions (into air, water, and soil) have been severely restricted for some years. Products made under these stringent standards will automatically score high in this area. North American manufacturers who comply with European emissions standards can be examples for others who now meet only minimum requirements. The United States Environmental Protection Agency (EPA) and Environment Canada have also set new, lower emissions limits for many pollutants, with deadlines for when they must be met. For example, by publication time, the use of CFCs in most foamed-plastic building products will have been discontinued. Recognition will be given for measures undertaken by manufacturers to improve primary and secondary production, such as redesigning the production process, substitution with safer solvents, dust controls, smoke stack particulate collectors (such as those which reduce contributors to acid rain), stack scrubbers, waste water treatment improvements, and in-plant solid and toxic waste reduction programs.

Few companies are willing to update to less polluting manufacturing methods before legislation requires it. Those which do sometimes face higher manufacturing costs. Making these efforts known can allow designers and consumers to make a conscious choice and to support these companies in their efforts.

Definition of our criteria: We give credit to products whose production is inherently low emission, for which in-plant measures have been made to reduce gaseous, liquid, and solid waste, as well as to reduce toxic waste which currently exceed regulations, or where significant improvements have been made in the past three years.

PACKING AND SHIPPING

A good deal of the energy use and solid waste associated with a product is incurred in packaging and shipping. This is particularly true for fragile and bulky items. Assembled furniture which is shipped long distance, for example, may be individually wrapped in plastic with a wood pallet under it and a reinforced cardboard carton on top. The packaging and shipping costs may be a substantial part of the product's total cost and a large part of these packing materials may be sent to a landfill.

Products with reduced impact due to their simplified packaging and shipping requirements are recognized under the following criteria.

MINIMUM, RECYCLED, RECYCLABLE PACKAGING

For most consumer products packaging has become a key marketing feature, but it generally does not play such an important role in building products. As a result, there is more scope for building products and furniture packaging to be designed for recycling or reuse. There are already small active networks for the recovery and reuse of skids, pallets, shipping blankets, and containers, and for recycling cardboard, plastic, and metal straps from shipping and packaging materials. The reuse of large plastic shipping bags is also being practiced. Under this criterion we give recognition for measures such as reduced packaging (especially the avoidance of shrink wrap and plastic foams), and the use of pallets or returnable bags, and recycled cardboard.

An additional factor for recycling materials which we cannot fully account for in our evaluation is that recycling facilities should be available locally. Because this cannot be determined, a recyclable packaging claim will be accepted if the supplier accepts returned packaging or if recycling facilities are common in many regions.

Definition of our criteria: We give credit to products for which packaging is either minimal, durable, and reusable, returnable, or fully recyclable (with widespread recycling facilities throughout North America).

MINIMUM TRANSPORT ENERGY

As well as the embodied energy from the production process, energy consumed during shipping must be considered in the evaluation of a product's environmental impact. By supporting and purchasing local manufacturers' products we can reduce that impact.

The main considerations are the volume which the product occupies, its weight, the distance from producer to market, and the type of transport. (Shipping by boat is several times more efficient than shipping by land because it takes less fuel per ton to move a ship).

Definition of our criteria: We give credit to products for which transportation impacts are reduced by an alternative shipping format (knock-down or dry-mixed), or by a company policy of regional production.

Three important stages in the life cycle of a product where health risks become a major concern are during its installation, its long-term use, and its maintenance. As we now know, indoor air pollution can only be partially resolved by mechanical means like exhaust ventilation. With air pollution from interior materials, the best method of reducing exposure during installation, use and maintenance is to reduce the sources of pollution.

The benefits of healthy working and living spaces are obvious: improved well-being and productivity, fewer complaints, and reduced incidence of illness. These benefits can easily offset higher initial material costs.

Health hazards are a very real concern for tradespeople. Solvents are in a highly volatile state before materials cure. Many materials such as adhesives and coatings are toxic or irritating. Fibrous materials and hazardous dusts from sanding, cutting, and handling also create health risks. Repeated exposure to these during installation can lead to such long-term health complications as respiratory disease, skin problems and chemical sensitization (with its allergy-like symptoms). The use of safer, low toxicity products, and preventive work practices are responsible ways to combat the risk of chronic illnesses in exposed tradespeople.

Exposure to toxic cleanup materials (mostly solvents) and irresponsible disposal of cleanup waste are also a concern in construction. Low toxicity materials tend to have safer cleanup procedures and less hazardous waste.

Definition of our criteria: We give credit to products which can be installed safely without special protective equipment and which do not produce significant hazardous or difficult-to-dispose-of waste.

Volatile organic compounds (VOCs) have become common emissions from many building and furnishing products. They are released either by the material itself or by the solvents, adhesives, stain and wrinkle resistance treatments, finishes, fire retardants, moth-proofings, fungicides, sealers, and waxes which are used in its manufacture or application. These offgas as the materials cure or deteriorate, usually producing a mixture of irritating gases. The use of chemically stable materials, and solvent-reduced paints, varnishes, and adhesives will greatly reduce the amounts and duration of the emitted VOCs.

Dusts are also generated by indoor materials during use due to their physical deterioration or their ability to trap soil. Wall, floor, and ceiling coverings, and liquid finishes are the major concerns for both gases and dusts by virtue of the large area they cover. These can continue emitting gases and releasing dusts long after they are installed.

Definition of our criteria: We give credit to products which are made from chemically stable materials, such as metals, ceramics, and other minerals. And also to those which have been specifically formulated, stabilized, or coated to reduce dusts, and to restrict total volatile emissions to a half-life period of one week or less after initial curing (that is, volatile emissions decline by half in one week).

A product's useful longevity is an important factor in evaluating its environmental impact. Durability limits the cycle of production-shipping-installation-removal and disposal. Recognition will be given for those products which typically outlast most other materials that might be chosen for the same application.

Definition of our criteria: We give credit to products which have above average life expectancy using standard durability testing methods for each category of material.

Easy maintenance must be considered when selecting a finish. Many common commercial cleaning agents are toxic and must be disposed of as hazardous waste. During use they are either caustic or emit gases, exposing the user and those in the building to health risks. Materials which do not require dry cleaning or special in-place cleaning, but which can be cleaned with low toxicity alternatives such as vinegar, borax, baking soda, and soap, are safer for the user and do not create disposal problems.

Definition of our criteria: We give credit to products which can be maintained, under normal use, without using specialized, in-place cleaning equipment or materials, particularly solvent-based, phosphate-based, and other high impact materials.

RESOURCE RECOVERY

The final stages in a building product's life cycle occur after its first use is over. At this stage it may still be reusable, or it may be recyclable. If it is neither, then it becomes either common waste or difficult-to-dispose-of waste. One difficult-to-dispose-of waste material is gypsum, which is severely restricted in many landfills due to its ability to produce hydrogen sulfide gas when buried.

Very few building materials made today are likely to become hazardous waste, though some lumber treated for rot resistance is hazardous. Demolition materials containing lead-based paints and asbestos from older buildings are also hazardous waste.

In the past, direct salvage for reuse has been a very minor source of building materials, but it may become a growing option. Antique and semi-antique doors, decorative millwork, cabinetry, and glass have great value in this era of eclectic "retro" taste. There is also growing respect for the quality of materials and craftsmanship of some of these older items. Materials or systems which are inherently reusable therefore have an advantage. Items are generally made available for reuse through salvage sales and by specialized salvage and demolition contractors, which can be found in classified advertising listings.

Definition of our criteria: We give credit to products which can be easily and practically removed for reuse, without major loss of quality.

"Recyclability" is defined as the extent to which a post-consumer material is returnable for reprocessing into the same material. "Conversion" refers to a post-consumer material converted into a lower value material. The two are quite distinct, though often misused. Recycling is more desirable, as it extends the raw material's usefulness. The extent to which a material may be recycled also depends on the presence of actual recycling facilities, at the time and in the region. Currently, recycling depots for paper, cardboard, container glass, some plastics, and most metals are available in larger communities in the U.S. and Canada. In some centers there are also recycling facilities for waste gypsum board from construction.

Definition of our criteria: We give credit to products which can be practically recycled into the same product, and for which there are actual recycling facilities operating in North America. Conversion of a product into another lower value material, though also worthy of merit and noted, does not qualify as recyclable.

SOCIAL AND ETHICAL ISSUES

More complex aspects arise when we consider social and ethical issues. We offer a general symbol for these issues, together with opportunities for more specific statements by manufacturers.

This part has a broader scope than the other aspects of manufacturing, and it provides an opportunity for manufacturers to make statements about their handling of the social issues. Further details are given in each individual Manufacturer's Statement. This area encompasses, but is not limited to, such issues as:
- Remuneration, benefits, and standard of living policies for employees;
- Working conditions standards beyond minimum requirements;
- Support and reintegration programs for employees affected by plant closures, downsizing, mechanization, or computerization;
- Social support for workers' families (daycare, education, relocation, pension);
- Cooperative or profit-sharing business incentives and job creation;
- Involvement in nonprofit charity organizations and community outreach programs;
- Equal employment opportunity for women, visible minorities, homosexuals, disabled people;
- Noninvolvement in military contracts;

• Noninvolvement in nuclear power;
• Equitable international trade practices.

Definition of our criteria: There are no clearly defined standards for this area, but respondents must list at least two innovative practices or policies to receive a positive rating.

This category allows manufacturers to offer information about their funding of research and education programs. Further information is given in the individual Manufacturer's Statement.

Definition of our criteria: There are no clearly defined standards in this area, but respondents must list the proportion of annual budget allocated to research and development, and any other specific support for independent research and education programs. Product research and promotion do not qualify.

CASE SCENARIOS:
MAKING CHOICES

We have prepared two case scenarios of the initial planning and information gathering stages of a residential and an office renovation. These "planning sessions" reveal the full implications of the choices we make regarding building materials, rather than considering only cost or fashion. Although cost and fashion are important, we can no longer afford to make them the prime considerations.

Making better choices for our own health and the global environment does make a difference, in our local communities and around the world. As consumers, designers, and builders begin to demand healthier materials for their homes and offices, industries will respond by producing more innovative and responsible interior finishing materials.

The following examples demonstrate a way of voicing concerns and defining criteria when planning new projects or renovations. These scenarios have been simplified to show the decision making process as clearly as possible. However, any project requires many more small decisions than we have included here.

The products mentioned are discussed in more detail in their respective sections. Desired properties for a particular product, such as low toxicity or durability, may be quickly checked on the Product Reports, where the symbol will be highlighted if it fulfills the criteria for that area.

As well, a summary table of materials is included at the end of each scenario. Such a table may be devised when planning a project, to help pinpoint concerns (e.g. environment, health, energy impact), and identify and weigh the benefits of the alternatives.

CASE SCENARIO: RESIDENTIAL

To illustrate the extent to which even minor construction can affect both the environment and our health, we have written a fictional conversation between a homeowner and a contractor. They are discussing a proposed interior home renovation, which will involve insulating the walls and ceiling, installing wallboard and a floor covering, and painting.

Homeowner: In the last few years, we've really made an effort to minimize the waste we produce. For this renovation we'd like to buy products that are recycled or that can be reused, and avoid materials that might affect our health. I've heard about a recently developed loose insulation made from recycled newspapers. It seems like a good choice for the environment, but is it as energy efficient as fiberglass?

Contractor: I've used cellulose fiber insulation on a few jobs this year. It insulates just as well as fiberglass, and isn't as hazardous to handle. It does have low toxicity additives to improve fire retardance, and protect against insects and fungus growth.

Homeowner: That sounds like a good product for everyone involved. Do you have any suggestions for the walls?

Contractor: There's a new type of wallboard that uses recycled newspapers mixed with gypsum. The gypsum fiberboard is stronger than conventional gypsum wallboard. It's also easier to install and prepare for painting because it has straighter edges and a harder surface.

Homeowner: I did some checking into paints, and I think I'd like to use a low toxicity type instead of conventional latex paint. No one in the family has allergies, so I don't think an all-natural paint is necessary.

Contractor: The low toxicity type is quite common now, and the lower solvent and biocide content is safer for my workers. Have you thought about flooring yet?

Homeowner: Yes. I've decided to buy high quality wool carpeting, since its production has less environmental impact than nylon, and it's durable. But I would like suggestions for undercushion and subflooring.

Contractor: Well, for years we put down carpet, usually nylon, over foam undercushion, with particle board underlayment. Although particle board is inexpensive, it releases formaldehyde, and the dust from cutting it is very irritating.

Homeowner: Is there a better alternative to particle board?

Contractor: Low density wood fiberboard made from recycled newspapers is safer to install and healthier for your family.

Homeowner: Good. Now what about foam cushioning? Another reason I chose wool carpeting is that it's biodegradable. I know that used synthetic carpets and cushions are usually simply dumped in landfills.

Contractor: Felt-type undercushions made from chopped waste fibers are a better choice. And, if we use nail strips to install the carpet instead of adhesives, it can be removed some day, and we avoid toxic emissions. A good wool carpet could be reused in your rec room in 10 to 15 years.

Homeowner: We probably would do that. That answers a lot of my questions, and I think we've found some alternatives that are better for us and for the environment.

MATERIAL CHOICES

	Conventional	Alternative	Merits of Alternative
Walls/ Ceiling:	Fiberglass batt insulation	Cellulose fiber insulation	Uses recycled newsprint Formaldehyde-free, low risk fiber
	Gypsum wallboard	Gypsum fiberboard	Recycled content Less filler needed Less dust produced
	Latex or acrylic/latex paint	All-natural paint	Traditional, safe ingredients, little or no petroleum products
		Low toxicity paint	Reduced solvents, biocides & other toxic contents
		Low biocide paint	Reduced biocides Safe handling
Flooring:	Particle board underlayment	Low density fiberboard	Uses recycled newsprint Formaldehyde-free
	Foamed plastic undercushion	Fiber undercushion (from waste textile fiber)	Recycled content Very low odor
	Synthetic fiber carpet (nylon, polyester) with latex bonding	Natural fiber carpet (wool, sisal, coir)	Renewable fiber Biodegradable
		Synthetic fiber carpet with fusion bonded backing or low emissions natural latex	Low odor construction

CASE SCENARIO: COMMERCIAL

To illustrate the differences between commercial and residential interiors, our second example depicts a company representative and an interior designer discussing requirements for a small office renovation, including a new reception area, work stations, and flooring in the lunch room. It emphasizes the durability and flexibility required for office interiors, and the major role played by furnishings.

Representative: We'd like to project our corporate image in the reception area, while keeping in mind the environmental impact of the materials and the overall cost. Will we be able to achieve that image with products other than exotic woods and marble, perhaps even using materials with some recycled content?

Interior designer: There are several products that use waste materials, such as some glass block products and ceramic tiles with recycled glass content.

As for tropical wood paneling or furniture, there are excellent alternatives that use reconstituted low grade wood. It is dyed, and can reproduce exotic colors and grain patterning with absolute consistency. These veneers also create less waste and need less finishing, keeping costs down.

Another little known material is cement based, and uses recycled cellulose and mineral fillers. This is every designer's dream material. It can be precast to virtually any form, and can be used for paneling, moulding, countertops, furniture, flooring, tiles, and more. Textures, colors, and additive aggregates are limitless, so we can have complete control over the material.

Representative: These certainly sound like the kind of options we had hoped for. The other item that needs addressing is the lunch room floor. It must be tough, and easily cleaned.

Interior designer: A good substitute for vinyl flooring is cork. It is naturally resilient, nonstatic, and water resistant, and it will reduce the noise level of the area. It can be installed as easily as vinyl sheet goods, although low toxicity adhesives should be used to keep offgassing to a minimum. As you know, it's difficult to contain indoor air pollution because of the air recirculating systems in office buildings.

Another option is natural linoleum, which has made a comeback. It's made from linseed oil and cork, lasts a very long time, is low maintenance, and will naturally resist bacteria, which cause sour odors. Properly sealed, a low toxicity cleaner is all that's needed for both cork and linoleum floors.

Representative: I think either one would be suitable. Now, in the general work area we want high quality, durable carpeting. Are there many choices?

Interior designer: If you're concerned with air quality, an efficient way to minimize carpeting adhesive odors is to install carpet tiles. Some use a grid of glued-down tiles that holds the others in place, and others have a releasable, low toxicity adhesive that allows worn tiles to be moved or replaced.

Representative: Good, because that was another prerequisite we have for this area—flexibility. That's our most important concern for the furniture. We expect the company structure to change over the next few years, and we may eventually need to move to other premises. We want the work stations and storage to be able to be moved around and added to as our needs change.

Interior designer: There are many furniture systems available. A recent addition to some lines is an option to buy refurbished pieces that are compatible with the original components. Some choices for materials are reconstituted wood products, domestic hardwoods, and powder-coated metals.

Modular storage/wall systems are available now that eliminate the need to build fixed walls and built-in cabinets, allowing for even greater flexibility for your whole work area, and less waste when changes are needed. That can reduce both your long-term costs and the environmental impact.

MATERIAL CHOICES

	Conventional	Alternative	Merits of Alternative
Reception Area/ Walls:	Glass sheet	Glass block	Recycled content, durable
	Tropical or domestic	Reconstituted veneer	Conserves wood Uses low grade woods
Flooring:	Marble/granite	Cellulose/cement composite	Recycled content Reduces shipping costs Less breakage
		Recycled glass tile	Recycled content Durable
Lunch Room/ Flooring:	Vinyl sheet flooring	Cork flooring	Renewable, natural contents
		Natural linoleum	Renewable, natural contents Naturally antibacterial
General Work area/ Flooring:	Glued-down broadloom	Releasable carpet tile, or other releasable system	Can be relocated to extend life Low emission adhesive
Furniture:	Traditional furniture	Systems furniture	Flexible, can be reconfigured and reused
	Built-in cabinets	Demountable storage walls	Flexible, can be reconfigured and reused

HOW TO USE THIS BOOK

We have designed this sourcebook to provide diverse information, from the general to the very specific. For easy and logical access, the sections have been divided into categories according to material use.

THE PRODUCTS

At the beginning of each section is a general discussion of that group of products, for example, wall coverings. The discussion covers the industry's standards and practices, including manufacturing and composition of the products, installation, maintenance, and resource recovery.

THE SYMBOLS

Before using the Product Reports, you should first read *About the Product Reports,* pages 6–14, to familiarize yourself with the visual symbols used, and the criteria that a product must meet in order to obtain recognition. The symbols on each Product Report provide at-a-glance information about the generic material, the specific product, and the company.

THE PRODUCT REPORTS

You can quickly find a specific product under its general category and material type (for example, *Wall coverings, fiber*). A description of the product includes its composition, applications, and available colors and sizes. Related Products include finishes and installation materials designed to be used with the listed product.

The manufacturers' names and addresses are provided and, while regional dealers and agents are not, a telephone number is given for further information on regional suppliers. For those industries where mail-order is an important means of distribution, the mail-order distributor is shown as the main contact. This occurs primarily for foreign products and small cottage industries.

Notes by the authors follow some Product Reports, where it was felt clarification was needed.

THE MANUFACTURER'S STATEMENT: This area in the Product Reports is reserved exclusively for comments by the manufacturer. The information published in this area has been reformatted and abbreviated, but it remains essentially unchanged from the manufacturer's original response. Where a response to our questionnaire was incomplete or conflicting, information was assembled from authorized technical specifications, research papers on generic materials and industry practices, and other information available in the public domain.

MEASUREMENTS

The measurements of materials have been given in the units used by the manufacturer. A metric/imperial conversion table may be found on page 242.

THERMAL INSULATION

Thermal insulations serve to keep heat inside in winter and reduce overheating in summer. They do this by trapping air or other gases in cells which transfer the heat slowly, since gases are not good conductors. The measure of the effectiveness of an insulation, known as its "R-value" (resistance value), is the degree to which it holds the captured air or gases and prevents heat from moving through it.

Insulations play an important role in reducing the environmental impact of buildings, because they reduce the energy required to keep buildings warm in winter and cool in summer. About one third of the total annual energy consumption in the U.S. is for the heating and cooling of buildings. The environmental consequences in terms of air and water pollution, resource depletion, and health risks to people and wildlife are enormous. Simply improving the insulation of buildings in North America would offset the demand for millions of barrels of oil, thousands of acres of land flooded for hydroelectricity, and all proposed nuclear power developments, with their consequent waste problems.

Unfortunately, most insulations are hazardous to handle, and are produced by industries with poor environmental records. Two significant exceptions, examined in this guide, are insulations made with shredded, post-consumer, waste paper fiber, and low toxicity foamed silicates. The paper fiber insulation uses a waste material which is in oversupply, and is safer to handle than other insulations. The foamed silicate is nonflammable, contains no hazardous mineral fibers and has a very low odor compared to other insulations.

MATERIALS

Many materials have traditionally been used for building insulations, including sawdust, straw, paper, bark, shredded plant fibers, and volcanic pumice. The majority of insulation today is made from spun glass fibers or foamed plastics.

FIBER BATT INSULATION

GLASS FIBER BATTS AND LOOSEFILL

Glass fibers are produced by melting natural, low quality glass materials (silica sand and limestone) in a furnace, and extruding the molten material into fibers. Although the raw materials for this process are widely available, the conversion process utilizes large amounts of energy. In addition, there are only a few manufacturing plants, usually situated a long distance from their markets, and shipping is costly due to the bulk of the final product, even when compressed. At present, there are not any recycled materials in fiberglass insulation, though this is technically feasible.

The spun fibers are either formed into batts (blankets) bound together with a polymer, such as phenolic resin, or are chopped into small clumps and used as loosefill material, containing mineral oil or silicone oil to control dust and keep it from separating. The fiberglass is naturally yellow, but a pink dye is added by some manufacturers to differentiate their product from others. Although the resins, dyes, and oils form a small part of the overall product, they are manufactured entirely from petroleum and coal byproducts, producing air and water pollution, health risks to workers, and toxic waste at all stages of production.

Fiberglass batts are available covered with kraft or asphalt-treated paper, or aluminum foil. Some form of vapor barrier over insulation is required in any type of construction, and is not unique to fiberglass insulation. Polyethylene sheet, usually applied during construction, has the lowest environmental impact of any of the materials for this purpose, because it requires very small amounts of a simple plastic made from natural gas. It is also the most effective, commonly used vapor barrier. Foil coverings add to the insulating value but are difficult to join and seal effectively. Where asphalt-treated paper is used as a facing, many more resources are consumed in its fabrication, and there may be objectionable odors that last a long time.

ROCK WOOL BATTS AND LOOSEFILL

Chopped mineral fiber (rock wool) insulation has similar properties to fiberglass but is made from mineral mining waste, with the benefit of reducing solid waste from another industry. The energy required to make it is similar to that for fiberglass, and it is also a health hazard during handling. Depending on the type of mineral waste, the air pollution from manufacturing and the risk to workers may be higher than for fiberglass, due to traces of heavy metals (cadmium and lead) from the raw material. Trace mineral elements vary from region to region. Loosefill rock wool is found predominantly in older buildings, though it is being manufactured today and is gaining in popularity.

BOARD INSULATION

FIBERGLASS

Fiberglass is also available in the form of board insulation. It has similar characteristics to batts and loosefill (see above), but with a higher density. It is a yellow, fibrous board available in medium to high density types and is usually used as exterior insulation, basement insulation, and for heating ducts.

Wood fiberboard consists of fibers and binding resins, and is similar to the recycled cellulose fiberboards used for exterior insulating panels. It may also contain low toxicity additives, such as flame retardants and moisture inhibitors, or be saturated with asphalt. The asphalt-saturated variety produces odors which may be objectionable to some peole. All wood fiber and cellulose insulating boards are light brown to grey in color, and lack the strength for structural uses. The recycled cellulose products utilize post-consumer waste paper as raw material, thereby reducing the amount entering landfills.

POLYSTYRENE

Both "extruded" and "expanded" polystyrene foam insulations are available. Expanded polystyrene (Styrofoam) is composed of small beads fused together, and expanded with a hydrocarbon (pentane). The board's R-value is lower that of the extruded variety. Expanded polystyrene board (Styrofoam) is usually white and brittle, and its bead structure can be clearly seen. It is often used as basement insulation or as a base for acrylic stucco exterior walls.

The extruded type contains polystyrene which has been foamed with a chlorofluorocarbon (CFC) or a hydrochlorofluorocarbon (HCFC), and the air and gas are captured in closed cells within the board. Extruded polystyrene is usually blue or pink, and is quite strong. It is sometimes used under concrete slabs or in basements.

Although fire retardants are added, both insulation types will still burn and emit toxic fumes, such as carbon monoxide. They also decompose under ultraviolet light and release the gases into the air space. Building codes require that polystyrene insulations be covered with a noncombustible finish.

POLYURETHANE (POLYISOCYANURATE) FOAM

Polyurethane can be a flexible foam for use in upholstery, or a rigid foam for insulation, depending on the isocyanate used. The ingredients (polyol resins, isocyanates and an amine catalyst) are expanded by a blowing agent, usually a CFC or HCFC gas. This process captures the gases in the insulation's cells, giving it a higher R-value than most other insulations. However, over time the gases will escape, reducing the insulation ability and contaminating the air. Although fire retardants are added, polyurethane is somewhat flammable and produces lethal gases as it burns—carbon monoxide, and oxides of nitrogen and hydrogen cyanide.

Polyurethane foams are used for insulating sheathing (exterior), and as a foamed-in-place insulation for some manufactured wood building panels. It is also available as a consumer spray product (expanding foam), used to fill gaps around windows and doors when weather sealing homes.

CELLULAR GLASS

Used primarily in commercial situations, cellular glass is nonflammable and moisture resistant. It consists of foamed glass, without fillers or binders. Carbon monoxide and hydrogen sulfide are trapped in the foam cells, and once installed, are not expelled.

The health risks from dust during installation and when cutting are serious, and precautions should be taken. Otherwise it is a very safe and odor-free material.

LOOSEFILL INSULATION

CELLULOSE

Cellulose fiber insulation is derived from fiberized (chopped), post-consumer newspaper. Waste newspapers, currently being stockpiled, and poor demand for recycled newsprint products have resulted in storage problems, so that the abundant "raw material" is expected to be available for a long time.

The product contains low toxicity additives to improve fire retardancy, reduce settling, and protect against insects and fungi. It is generally blown into place for fast, economical installation and minimal handling.

Its environmental profile is excellent because it uses post-consumer waste which is currently a surplus material. Cellulose requires very little energy in its manufacture, though it may be shipped some distance from the manufacturing centers. Its bulk results in a large energy expenditure for transport.

Cellulose insulation's health profile is good as well, because it is made of nonhazardous fibers that contain no resin binders. Installation is safer than for glass and mineral fiber insulations. The residues of printing inks and the additives, however, are slightly toxic and produce an odor. Like any insulation, it must be thoroughly sealed into the cavity, so that fibers and odors are prevented from entering the interior air space.

VERMICULITE AND PERLITE

Vermiculite is a mineral that resembles mica, and contains both free and chemically bound water. Perlite is a natural silicate volcanic rock that is very dusty. These are poured in place, and are used primarily inside the cavities of concrete and masonry blocks, or in insulation plaster and lightweight concrete. Silicone oil is added to control the dust. Both are expanded by heat, are inherently flame retardant, and resistant to rot and termites. They produce hazardous dusts during handling, but are safe and odor-free once installed. Vermiculite and perlite are easily recovered for reuse.

POLYSTYRENE BEADS

Polystyrene beads are similar to expanded polystyrene rigid insulation, but are installed loose. Primarily for use in concrete and masonry blocks, the effects and health hazards are the same as for the board insulation. The material is easily recovered for reuse.

It is imperative to seal the cavity from the living space with all loosefill insulations, as the materials are partially reduced to a fine dust that can easily seep through cracks and contaminate the air.

FOAMED-IN-PLACE INSULATION

UREA-FORMALDEHYDE FOAM (UFFI)

UFFI was used primarily in existing houses in the 1970s, but was banned for most uses by 1980. Due to health concerns, in most areas the law now requires its presence to be disclosed to prospective house and building buyers. The gases released include formaldehyde, benzene, benzaldehyde, acetaldehyde, cresol, methylnaphthalene, acrolein, ammonia, and phenol. All are very toxic, and formaldehyde has the further effect of sensitizing people to other chemicals, with the potential to develop a wide range of symptoms and health complications.

Though UFFI does not produce gases any longer after several years, the dust is very irritating and makes thorough removal difficult. Consult your local health authority or Environmental Protection Agency (EPA) office for further information if there is UFFI in your building, particularly if renovations are anticipated.

SILICATE FOAM

A recent innovation, foamed-in-place silicate insulation is purported to be nontoxic, and at the very least is a low toxicity product. It contains magnesium oxychloride and sodium silicate, both inert materials, and the blowing agent is compressed air. A pink dye and detergent foaming agent are added that may exude odors, but the dye can be left out by special order. This product is recommended for concrete and concrete block insulation, though it is also used in wood frame buildings. It does take several days to cure and remains moist during that time, however, so should be used with caution in wood and gypsum board construction.

INSTALLATION METHODS AND HEALTH RISKS

BATTS AND LOOSEFILL

Batts are unwrapped and placed in frame spaces by hand, while loosefill is either poured in or blown in with a machine. Health risks for both types occur during manufacture and installation, when industry workers and tradespeople are exposed to high levels of the glass and mineral fibers. These fibers penetrate the skin, causing itchiness and serious irritation to the nose and eyes. Though not as dangerous as asbestos, inhaling the very small fibers is considered to be a high health risk that may result in respiratory diseases, including cancers. Proper protective clothing (gloves, long-sleeved shirt, pants, goggles, and respirator mask) can reduce the severity of exposure. Fibers will remain on equipment and clothes, and can further expose the worker or others who come in contact with them, unless they are thoroughly washed. Pouring and blowing methods release far more fiber than hand installations.

There is little health danger for the occupants of a finished interior with fiberglass, rock wool, or cellulose insulation, if the material is completely sealed from the air space with a polyethylene vapor barrier, foil barrier, or caulked and gasketed gypsum board system. Fiber leakage can occur, however, particularly in older buildings which may have gaps around windows, doors, and electrical outlets. This is less of a concern with fiberglass batts than with loosefill insulation. In the event of fire, fiberglass is inert, but the binders and facing materials will emit highly toxic gases.

If installation is carefully executed and the material has not been degraded by contact with moisture, fiberglass and rock wool batts can be removed and reused. This is rarely done, though, due to the difficulty of handling and storing such bulky material. There is currently no facility for recycling fiberglass or rock wool, though it could be chopped and reused as loosefill insulation.

BOARD INSULATION

Glass fiber and cellular glass boards release small amounts of hazardous fibers during handling and cutting, but protective clothing and respirator masks will provide protection. Rigid polystyrene and polyurethane are not hazardous to handle unless they are cut with power tools or a hot wire method, when they will release hazardous gases. Wood fiberboard is quite safe to handle, and is usually cut with a knife.

None of the rigid board insulations is currently recyclable, though all can be reused if removed carefully. Except for the cellular glass, which is brittle, all rigid insulation waste could potentially be chopped and used as loosefill material, if the equipment were available.

BLOWN-IN AND BLOWN-IN BATT (BIB)

Fiberglass, rock wool, and cellulose loosefill insulations can also be blown into wall cavities in new construction using a method called blown-in batt. One side of the wall is left open but covered with a flexible screen material. A blower forces the loosefill insulation through a hose into the cavity until it is full, and the vapor barrier and wallboard are then installed. It is an effective system for walls, and can make use of loosefill insulations with recycled contents. However, blown-in insulations increase fiber exposure dramatically over hand installed batts because fans disturb the material.

FOAMED-IN-PLACE INSULATION

Polyurethane insulation, foamed in place, is moderately hazardous to handle, expensive, and very hazardous in a fire. Foamed silicate, as described above, is much safer to handle and nonflammable, though it is also expensive. These are the only foamed insulations usually available.

PRODUCT
Cellulose fiber insulation

MANUFACTURER
Can-Cell Industries Inc.
(National Cellulose)
16355 130th Ave.
Edmonton, AB T5V 1K5
Tel (403) 447-1255   Fax (403) 447-1034

DESCRIPTION
WEATHERSHIELD INSULATION: 100% recycled fiberized paper, of which 80% of product weight is post-consumer newsprint, treated with borax and boric acid as a flame retardant and insect repellent. R-value of 3.6 per inch.

Product has low density, with overall low energy consumption in manufacture. It can be raked in place, or blown in pneumatically.

MANUFACTURER'S STATEMENT
Weathershield insulation has been awarded Environment Canada's designation as an Environmental Choice product. It has met the criteria outlined for insulation materials and carries the Ecologo label.

State-of-the-art technology has recently been implemented that addresses energy efficiency, an air filtering system, and fiberization equipment, which increases overall production efficiency.

Can-Cell has recently completed two insulation research projects in conjunction with the Alberta provincial government, under the Innovative Housing Program.

PRODUCTION

RECYCLED CONTENT

IN-PLANT ENERGY EFFICIENCY & RECYCLING

LOW EMISSIONS PLANT

PACKING / SHIPPING

INSTALLATION / USE

MINIMUM INSTALLATION HAZARDS

SEE PAGE 25

LOW TOXIC EMISSIONS IN USE

RESOURCE RECOVERY

REUSABLE, SALVAGEABLE

OTHER

RESEARCH & EDUCATION PROGRAMS

RECYCLED
CONTENT

PRODUCTION

PACKING / SHIPPING

MINIMUM
INSTALLATION
HAZARDS

SEE PAGE 25

LOW TOXIC
EMISSIONS
IN USE

INSTALLATION / USE

RESOURCE RECOVERY

OTHER

PRODUCT
Cellulose fiber insulation

MANUFACTURER

International Cellulose Corp.
P.O. Box 450006
12315 Robin Blvd.
Houston, TX 77245-0006
Tel (713) 433-6701, (800) 444-1252
Fax (713) 433-2029

Can-Cell Industries Inc.
(National Cellulose)
16355 130th Ave.
Edmonton, AB T5V 1K5
Tel (403) 447-1255
Fax (403) 447-1034

DESCRIPTION

CELBAR (WALLBAR available through Canadian manufacturer only): 100% fiberized post-consumer newsprint, treated with borax and boric acid as flame retardants. Pneumatically spray-applied in wall and ceiling cavities. Provides sound and thermal properties, without allowing sound leaks or air infiltration. Complete coverage and sealing is achieved. Provides R-value of 3.8 per inch.

K-13: Spray-applied cellulose insulation designed to remain exposed. 100% recycled fiberized paper, of which 80% of product weight is post-consumer newsprint. Borax and boric acid are used to impart flame retardant properties, instead of ammonium sulfide, which breaks down into sulfuric acid on contact with moisture and undermines the integrity of buildings with metal structures. Has Class 1 fire prevention rating.

Provides acoustical and reverberation absorption properties. Aids in condensation control in damp areas, such as indoor swimming pools.

The binding agent is a patented, nontoxic, odorless adhesive, and is combined with the fibers during application. Product is spray-applied with licensed fiber machines and nozzles for control of the fiber/binder ratio and resultant fine texture.

Available in 5 standard colors, with the ability to be matched to Pantone Matching System, or any other universal tinting system on request. The texture can be altered by rolling, tamping, painting, or overspraying.

K-13 "FC" CEILING SYSTEM: Designed for exposed finished surface application to any substrate. Has acoustical, thermal, and light reflectivity properties. It reduces or eliminates concrete preparation steps.

Contains 100% recycled paper that has not been printed, such as photographic paper, and mill waste. Color matching to any universal tinting system is available.

MANUFACTURER'S STATEMENT

The specifically engineered, nontoxic water-dispersion adhesive was developed to be a safe material. The products containing recycled newsprint were developed based on the long tradition of using newspaper as insulation in walls, and the ready availability of a recycled resource.

PRODUCT
Cellulose fiber insulation

MANUFACTURER
Louisiana Pacific
111 S.W. 5th Ave., Suite #4200
Portland, OR 97204
Tel (503) 221-0800   Fax (503) 796-0204

DESCRIPTION
NATURE GUARD: 100% recycled paper loosefill insulation with R-value of 3.6 per inch. Blown into place, it seals around irregular objects, resulting in less air infiltration and greater sound absorption. Fibers stay uniform and fluffy.

Nontoxic, nonflammable and safe during installation.

MANUFACTURER'S STATEMENT
Louisiana Pacific is working to produce innovative, affordable products that are environmentally sound. Their goal is to provide products that are better, more cost effective, and easier to install than conventional materials. The company is very concerned with the rising costs of house building and is actively promoting awareness and encouragement of affordable housing.

The company has recently invested $75 million to upgrade pollution controls at their pulp mill in California, $65 million of which was for a new recovery boiler, resulting in significant improvements in air and water emissions (after a settlement with the Environmental Protection Agency for $3 million for past violations of waste water discharge limits).

They anticipate depending less on forests in the future, using fiber from drip-irrigated plantations which produce a harvestable crop of trees every seven years. They have planted Italian poplar and eucalyptus, and can control the site selection for these plantations so they are close to mills and markets.

PRODUCTION
RECYCLED CONTENT
IN-PLANT ENERGY EFFICIENCY & RECYCLING
LOW EMISSIONS PLANT

PACKING / SHIPPING

INSTALLATION / USE
MINIMUM INSTALLATION HAZARDS
SEE PAGE 25
LOW TOXIC EMISSIONS IN USE

RESOURCE RECOVERY
REUSABLE, SALVAGEABLE

OTHER
RESEARCH & EDUCATION PROGRAMS

RECYCLED CONTENT

PRODUCTION

PACKING / SHIPPING

INSTALLATION / USE

SEE PAGE 25

LOW TOXIC EMISSIONS IN USE

RESOURCE RECOVERY

OTHER

PRODUCT
Mix of cellulose and mineral fiber insulation

MANUFACTURER

American Sprayed Fibers
1550 E. 91st Dr.
Merrillville, IN 46410
Tel (219) 769-0180, (800) 824-2997

Able Mechanical Insulation Ltd.
8515 152nd St.
Surrey, BC V3S 3M9
Tel (604) 597-0100

Advance Insulation
19A Gino St.
Sudbury, ON P3E 4K2
Tel (705) 675-7408

DESCRIPTION

DENDAMIX: Fireproofing insulation consisting of blended, fiberized post-consumer newspaper and mineral fiber (slag wool) in a water-soluble liquid adhesive. Wet-spray applied. R-value of 3.25 per inch.

THERMAL-PRUF: Thermal insulation consisting of blended, fiberized post-consumer newspaper and mineral fiber (slag wool) in a water-soluble liquid adhesive. Wet-spray applied. R-value of 3.8 per inch.

THERMAL-PRUF TIGER BRAND: Thermal insulation consisting of 100% fiberized mineral slag in a water-soluble liquid adhesive. Wet-spray applied. R-value of 3.8 per inch.

SOUND-PRUF: Acoustical insulation consisting of blended, fiberized post-consumer newspaper and mineral fiber (slag wool) in a water-soluble liquid adhesive. Wet-spray applied. R-value of 3.8 per inch.

MANUFACTURER'S STATEMENT

The heat-retardant adhesive was specifically developed to encapsulate asbestos. The products containing recycled newsprint and waste mineral slag were developed for the asbestos replacement market.

All products are nonhazardous and nontoxic, containing no asbestos or fiberglass.

NOTE

Slag wool is also a potentially hazardous fiber and should not be installed where deterioration may expose building occupants to it.

Cotton fiber insulation

MANUFACTURER

Cotton Unlimited, Inc.
Old Mill Rd.
Post, Texas 79356
Tel (806) 495-3501   Fax (806) 495-3502

DESCRIPTION

INSULCOT: Made of 75% below grade cotton or textile industry waste by-product, and 25% polyester fibers. Product is treated with a nonhazardous borax-based flame retardant that is absorbed into the fiber.

Produced in batts and rolls, for easy installation. Meets Class I fire rating standards for habitat insulation (similar to fiberglass).

The product is manufactured in various thicknesses, ranging from 3" to 8", achieving corresponding R-values of R-11 to R-30. These values are similar to that of fiberglass, but without the hazard during handling and installation.

MANUFACTURER'S STATEMENT

Manufactured by modified standard textile equipment, with low utility costs and no hazardous materials emitted into the air.

Another plant is presently being built to step up production, and allow greater availability.

PRODUCTION

SUSTAINABLY ACQUIRED, OR RENEWABLE RESOURCE

IN-PLANT ENERGY EFFICIENCY & RECYCLING

LOW EMISSIONS PLANT

PACKING / SHIPPING

INSTALLATION / USE

MINIMUM INSTALLATION HAZARDS

LOW TOXIC EMISSIONS IN USE

RESOURCE RECOVERY

REUSABLE, SALVAGEABLE

OTHER

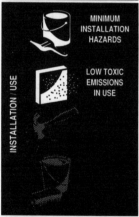

MINIMUM
INSTALLATION
HAZARDS

LOW TOXIC
EMISSIONS
IN USE

PRODUCT
Cementitious foamed-in-place insulation

MANUFACTURER

Air Krete
(division of Nordic Builders)
162 N. Sierra Ct.
Gilbert, AZ 85234
Tel (602) 892-0603

I.F. Industries
1426 Wallace Rd.
Oakville, ON L6L 2Y2
Tel (416) 827-6358
Fax (416) 827-9342

DESCRIPTION

AIR KRETE: A foamed-in-place mixture of sodium silicate, magnesium oxychloride, modified detergent, dye, and air. Air Krete is ultralight and inorganic. It contains no formaldehyde, asbestos, or chlorofluorocarbons and has the appearance of cement plaster. It is expanded with compressed air, and is fully expanded upon leaving the application equipment nozzle.

Air Krete is nonflammable, nontoxic, and has an R-value of 3.9 per inch. It is an effective noise reduction barrier and lends itself to new as well as existing construction.

MANUFACTURER'S STATEMENT

Air Krete is nontoxic even for those who are chemically sensitive. There are no irritating fibers, no chemical gases, or vapors during or after installation. It is pest and moisture resistant, and ozone safe. Contains no asbestos, no formaldehyde, and no chlorofluorocarbons.

Air Krete is approved and used by public school systems, Housing and Urban Development (HUD), Army Corp of Engineers, Veterans Administration, National Bureau of Standards, food processing plants, industrial and residential construction.

INTERIOR CONSTRUCTION PANELS

MATERIALS

Until the 1940s, all buildings were constructed using small dimension materials like 1" × 8" boards and ½" × 2" wood lath. Today, we assume that standard building panels, usually 4' × 8', will be used. A broad range of panel types are available, from plywood and gypsum board, to fully prefabricated wall, floor, and roof sections complete with insulation. Because this book covers interior construction only, the panel types listed here are composite wood boards, wood fiberboards, and gypsum board.

COMPOSITE WOOD BOARDS

PLYWOOD

Plywood is made from thin sheets of wood veneer which has been peeled from logs, dried and pressed together into sheets, using glue between the layers. For the best quality products, the process requires high quality logs from trees with straight grain and few knots. Though plywood is a very efficient material due to its strength and light weight, the demand for quality logs for veneer means serious pressure on remaining old growth timber. Domestic plywood is nearly all exterior grade, and is made from fir, spruce, and pine. Imported plywood is usually interior grade, and made from lauan (Philippine mahogany). The stocks of quality lauan are threatened in the tropics due to unregulated cutting.

Two common glue types are used in plywood manufacturing: the phenol-formaldehyde "exterior grade" glues, and the urea-formaldehyde "interior grade" glues. The exterior glue is a more stable composition, which emits very little formaldehyde and is moisture resistant. It can be easily identified by the exterior grade markings on the board and by its dark brown, molasses-like appearance. Interior grade glue is less stable and emits more formaldehyde, particularly under humid conditions. It is difficult to recognize because it is a light cream color, very similar to many woods. Interior plywoods are used for veneered wall paneling and for door skins.

Formaldehyde emissions from all plywoods can be reduced by sealing surfaces with lacquer-type coatings (nitrocellulose lacquers), or with specially formulated, low toxicity sealers. (See FINISHES, pp. 143, 158, 168, 173, 174, for more information and products.)

CHIPBOARD (WAFERBOARD) AND ORIENTED STRAND BOARD (OSB)

These structural boards are made from low grade hardwoods such as aspen and poplar, or from common softwoods such as hemlock and fir. The wood is chipped into large, thin strips, and then pressed together with glue. This process uses the resources of fast-growing species efficiently, and boards can be made from waste products, such as logs unsuitable for sawmilling, which would otherwise be burned or pulped.

The glue is usually phenol-formaldehyde based due to the moisture resistance required for exterior rated products. Though the emissions from this glue are far less than those from interior products, the glue content may be 10% or more of the board weight, which makes it still a health concern.

Chipboard, or waferboard, and OSB are generally designed for structural uses, such as sheathing panels for wood frame walls and roofs. They are also used as web material (the vertical piece) in manufactured wood joists. These boards are good choices for exterior construction due to their resource efficiency, but should be used with caution indoors due to the high glue content. They give off a strong "phenolic" odor which persists for years.

PARTICLE BOARD

These materials are made from softwood dust or chips, which have been pressed into sheets using glues to give them strength and surface hardness. They offer a stable, smooth, and scratch-resistant surface.

Particle boards are, however, generally made with large amounts of urea-formaldehyde and phenol-formaldehyde based glues, which make handling and cutting a toxic exposure risk for industry workers. In fact, carpenters and cabinetmakers who work with these materials routinely develop sensitivity reactions after several years. The reactions include skin inflammation and respiratory distress. Furthermore, if the formaldehyde-based materials are not chemically stabilized or completely sealed by veneer or special sealers, they will offgas formaldehyde and other irritating gases, adding to indoor air quality problems. They also release odors from the softwoods used in their manufacture. These odors, particularly from red cedar products, are quite allergenic to some people.

Particle boards are usually classed as medium density fiberboards (MDF), which means that they are formed under pressure to make them harder than plywoods. They are not, however, as hard as hardboards and tempered hardboards. Some of the most common types are made from pine, hemlock, or red cedar dust with urea-formaldehyde glue. The glue may comprise over 15% of the material by weight. These products generally use waste wood from sawmills, which would otherwise be burned or sent to landfills. There is an obvious environmental merit for this resource use, but it must be weighed against the health risks.

The most common indoor use for particle board is for flooring underlayment, cabinets, and furniture. (See FURNITURE AND ACCESSORIES, p. 206, for more information.) In installations, such as carpet underlayment, where large amounts of particle board are in a building's airspace and covered by a porous material, formaldehyde emissions are the most serious. This is one of the most common causes of air quality complaints in new homes.

Dust from cutting this material is far more irritating than ordinary wood dust, due to its high glue content. Particle board should be worked with caution, using dust collectors (typically found only in fabricators' shops) and high quality respirator masks.

There are a few alternatives to conventional particle boards. Some are made with exterior glues which emit less formaldehyde (classed as "exterior" or "exposure one" materials); others are chemically stabilized to meet more stringent emissions standards (classed as "low emission" or "European standard" materials); and some are made using no formaldehyde-based glues at all (classed as "formaldehyde free" materials).

HIGH DENSITY HARDBOARD (WOOD FIBERBOARD)

Fiberboard is a common term for manufactured wood products made from wood which has been ground and then pressed, using high temperatures and sometimes glues, into a thin, very strong panel. These boards are typically only one eighth to one quarter of an inch thick, and are dark brown with a textured or glossy finish. A common trade name for this material is Masonite.

These products can be made without glues because wood contains lignin, a natural polymer resin which holds wood fibers together. Though the lignin in fiberboard can bind the fibers together, glues or other additives are often used to give the board special properties. For example, the surface-treated or tempered varieties, which have a very hard and glossy surface, may have chemical additives.

High density fiberboard manufacturing is an effective use of wood resources as it can use lower quality fiber to make a high quality product. Due to the small quantities of glue and the high temperature processes, indoor air pollutant emissions are much lower than for other types of manufactured wood products. Though the manufacturing process uses a great deal of energy, the product is very durable and offers high strength from a small amount of wood. This makes its overall environmental cost relatively low.

The most common interior uses for hardboard are as textured wall paneling, as faces for hollow core doors, as pegboard for utility areas, and as furniture and cabinet backs and drawer bottoms.

LOW DENSITY WOOD FIBERBOARD

Low density wood fiberboard refers to either lightweight board made from new wood fiber or board made from waste newspaper. Wood fiberboard made from new wood fiber is produced in the same way as high density fiberboard (see above), but under lower pressure so that it remains quite soft. This material is used for acoustic tiles and sound-absorbing partitions, as well as tackboards and backing for fabric-covered panels. It is often supplied with white primer on one side. This product uses low grade wood fiber effectively, and can be made without glue due to the natural lignin in the wood.

An alternative to wood fiberboard made from new wood, which fulfills many of the same functions and more, is a board made from compressed newspaper. It is made entirely from post-consumer waste paper, which is in serious oversupply in most areas due to the lack of mills able to receive it. Because most of the lignin has been removed in the papermaking process, the board must be bonded together with an adhesive. A low toxicity adhesive containing no formaldehyde is used, which is more expensive than conventional glues.

The health risks from this product are small, but it does contain substantial ink residues which may have a slight odor. Though the dust is not very irritating, high quality respirators should be worn while machine cutting.

There are many variants of this product, with different densities and surfaces, suitable for acoustic paneling, carpet underlayment, as a base for fabric-covered panels, as tackboard, and for many other purposes. It has a very low environmental cost as it is made in a low energy, low pollution process from a waste material which is, at present, a storage and disposal problem.

GYPSUM WALLBOARD

Gypsum wallboard was developed as a faster and less costly alternative to manual plastering methods. Before the development of gypsum board, most interior plastering was done using wood lath (thin strips of wood placed close together) as a base for plasters made from sand, cement, lime, and gypsum. Gypsum board, also called sheetrock or gyproc, is a factory-made panel of gypsum sandwiched between two layers of heavy paper. The gypsum also contains fibers, such as glass fiber, to increase the board's strength and fire resistance.

Gypsum is calcium sulfate, a naturally occurring mineral found in chalk-like deposits. It is mined extensively along the Canadian Atlantic coast, in the prairie regions of Canada and the U.S., and in the southern U.S. The mines are shallow, open pit operations with some significant environmental problems, particularly from gypsum-laden mine runoff, which is a hazard to aquatic life. When gypsum is mined, it contains water as part of its molecular structure, which limits its use potential. In order to increase its usefulness, it must first be processed. Raw gypsum is crushed and roasted in large ovens to evaporate the water, after which it will react and set when water is added. These processing steps use a great deal of energy, and produce dusts and the air pollutants nitrogen oxide and carbon monoxide.

Gypsum is also produced as a byproduct of other industrial processes, and is then called synthetic gypsum. Though this type of gypsum has not been produced in quantity in North America, it will become very important in the near future. The reason for this is that gypsum is produced in the most common process for capturing sulfur dioxide gases from the stacks of coal-burning electrical generating stations. If the gases are allowed to escape, they become major contributors to acid rain, which is devastating forests and lakes in the northeastern U.S. and eastern Canada. The process, called Flue Gas Desulfurization (FGD) is now mandated under the U.S. Clean Air Act and Canadian regulations.

Gypsum produced by this means is slightly different from that which is mined. It can contain concentrated impurities from the coal, including metals, which can be health hazards. A great deal of this type of gypsum is now used in Europe to make wallboard, but it must first be tested for safety because regional variation in coal composition and flue gas equipment can make important differences in the trace contents of the gypsum. No building products made from this type of gypsum were found in North America during the survey for this book, but it is likely that synthetic gypsum products will be available soon.

HEALTH CONSIDERATIONS

Gypsum is very safe from a toxicity standpoint, except that the dust, like most mineral material, is irritating and slightly hazardous to inhale. When powdered gypsum is handled before it has been mixed with water, it is mildly caustic to the skin. Pure gypsum, however, is an approved food additive, and is used in tofu (bean curd) manufacture.

The main health risk from gypsum board is the installation procedure, which involves handling joint fillers and sanding the joints. This process creates large amounts of fine dust, which is both a nuisance and a minor health risk, and introduces exposure to another group of materials used in the fillers. Though most gypsum board is installed with screws or nails, it may also be done with a construction adhesive, which is a toxic material containing hazardous solvents, such as xylene and toluene. Virtually all interior installations are also primed and painted. (See INSTALLATION MATERIALS and FINISHES for further discussion.)

RECYCLING

Gypsum is recyclable, if it is not contaminated, and recycling facilities are now available in many locations. The main incentive for recycling waste gypsum is that it causes environmental problems in landfills, particularly in wet climates. (See DISPOSAL, below.)

Gypsum recyclers will typically receive only gypsum board scrap which has not been painted. The paints, fillers, and sealers permeate both the paper backing and the gypsum, contaminating it so that it cannot be recycled. The recyclers remove the paper surfaces, crush the gypsum, and send it to a gypsum board plant to add to the gypsum for new board. Recycled gypsum may also be sold as a soil conditioner or for other uses. Currently about 90% of the gypsum can be recovered from construction scrap.

DISPOSAL

When gypsum from demolition or construction scrap is dumped with other materials it can create serious problems, especially in wet climates. Gypsum which is buried in moist landfills and cut off from air will support bacteria (called anaerobic, because they live without air), which are able to break down its chemical structure. These bacteria produce hydrogen sulfide gas, a strong smelling, dangerous gas that leaches into the groundwater and renders it acidic. When polluted water enters streams and lakes it is toxic to fish and aquatic plants.

Five per cent gypsum content is currently the maximum allowable in many landfills, and dump loads are checked for gypsum content in regions with restrictions. Because gypsum from demolition is contaminated with paints and other materials, it is not recyclable. (See RECYCLING, above.) It usually goes to landfills, or is dumped in the ocean in coastal areas. Both practices can cause serious environmental damage and must be reviewed.

PAPER IN GYPSUM BOARD

The paper used in gypsum board manufacture is generally a kraft paper mill product, though it may also be derived from recycled newspapers or have some recycled content. In a paper mill, wood fiber from low grade trees or sawmill waste is ground and digested, using sulfates and caustics, and then formed into paper. Some types are bleached, usually using hazardous chlorine, though only moderate amounts of bleaching are done to papers for gypsum board manufacturing.

All pulp and paper processing, as currently practiced in North America, has serious negative environmental consequences in the form of logging, air and water pollution from mills, and high energy use. One environmental problem is the sulfite wastes from digesting pulp. These wastes are burned in plant boilers, producing acid fallout and serious damage to forests, as well as health risks. Another problem is chlorine residue from bleaching, that is discharged into rivers, lakes, and oceans, where it forms toxic dioxins. Using recycled fiber or minimizing the highly processed and bleached paper content of gypsum boards is one of the few options currently available for reducing the impact of paper production.

WATER-RESISTANT GYPSUM BOARD

Additives are incorporated in gypsum board to increase its moisture resistance, for use in damp rooms and as a base for exterior finishes. These additives may contain chemicals to resist fungus contamination of the board under these conditions. Water-resistant products are generally made from a conventional board which is saturated, under pressure, with water-repellent silicone oil or paraffin wax derived from petroleum. The water-resistant agents add a petroleum-based component to the board, increase the energy required for manufacture, and render the material unfit for recycling. Though the water-resistant agents are low in toxicity, additives for retarding fungus growth are toxic.

FIBER GYPSUM

An important innovation in gypsum boards is a recent introduction called fiber gypsum. This is a wallboard with no paper facings, which incorporates wood fibers into the gypsum itself for reinforcement. It is a stronger product, requires less taping of joints, and avoids the environmental disadvantages of kraft paper.

Gypsum fiberboard contains paper from post-consumer newsprint, which has been fiberized and mixed with the gypsum before setting. It has a much harder surface than conventional gypsum board, and does not have the problems of tearing, bulging, and scratching that paper-faced products do. The overall recycled material content is approximately 15%, which gives fiber gypsum a lower production, energy, and pollution factor than other gypsum boards. Gypsum fiberboard also requires less joint compound for installation as the edges are accurately machined to fit closely together.

There are several methods for installing gypsum fiberboard. Joints can be filled in a two-step process using the joint compound provided by the manufacturer, which comes in a tube to fit a caulking gun. This process eliminates the need for taping, although standard methods of taping and filling can be used. Corners can be finished using conventional corner bead accessories and filling compounds. If the standard methods are employed, the process will be familiar to tradespeople. Low toxicity joint compounds can then be used where needed. (See INSTALLATION MATERIALS, pp. 103, 110, 111.)

MANUFACTURER
Alpi s.p.a.
Imported by

The Dean Company
P.O. Box 1239
Princeton, WV 24740
Tel (304) 425-8701, (800) 624-6153
Fax (304) 426-2452

The Dean Company
1586 Louise St.
Laval, PQ H7S 1E4
Tel (514) 667-5945

RELATED PRODUCTS
Adhesives, pp. 105–6, 119–22
Sealers, pp. 142–3, 166–74

DESCRIPTION

ALPILIGNUM: Wood grain patterned raw veneers from reconstituted obeche (ayous) wood, from strictly government-controlled forest concessions in Africa. Alpilignum is virtually defect-free and offers color and grain consistencies that reproduce most domestic and exotic wood species. Color consistency is controlled by a computerized dyeing process and identical grain patterns are produced with the use of molds.

Veneer dimensions: wood grains are available in sheets 25" wide × 111", 122" and 134" long × 1/39" thick. Bird's eye and burls are available in sheets 25" wide × 89" and 99" long × 1/45" thick. Various thicknesses are available ranging from 1/84" to 1/8".

Lumber dimensions: 1" to 3½" thick. (Lumber available only on a custom basis, for a minimum quantity.) Grain patterns include: flat cut grains (heart patterns), quartered grains, bird's eye patterns, burl patterns, geometric patterns. Special grain patterns and custom patterns are available for minimum quantities.

Dyes are water-soluble aniline dyes and available in a great variety of colors and shades. Vat dyeing ensures complete color penetration. The adhesive used in the manufacture of alpilignum is a special formulation of urea resin with a low formaldehyde content. Water-based wood glues and contact cement are recommended for bonding to the substrate.

ULTRA-FLEX: Alpilignum veneer with a 10 mil paper backing for ease of application to the substrate. For use on interior surfaces, such as store fixture design, tambour, partitions with wood core, kitchen and bath cabinets, audio speaker cabinets, and residential and contract furniture. Material comes presanded and flexed. Prefinishing and cutting to size are available for special orders.

Veneer dimensions: wood grains are available in sheets 48"–52" wide × 132", 120" and 110" long × 1/39" thick. Bird's eye and burls are available in sheets 48"–50" wide × 98" long × 1/45" thick. Various thicknesses are available ranging from 1/84" to 1/8".

A wide range of flexibility is obtained with 5 mil, 10 mil, and 20 mil paper backings. For custom orders, 5 mil or 20 mil paper backings, pressure-sensitive adhesive backing or cloth backing/clay-filled backing (with Class A fire rating) are available.

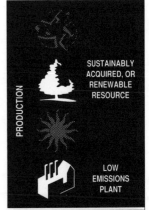

PRODUCTION

SUSTAINABLY ACQUIRED, OR RENEWABLE RESOURCE

LOW EMISSIONS PLANT

PACKING / SHIPPING

INSTALLATION / USE

MINIMUM INSTALLATION HAZARDS

LOW TOXIC EMISSIONS IN USE

RESOURCE RECOVERY

OTHER

Oil stain, linseed oil, or wax finishes are not recommended. An ultraviolet inhibitor is recommended in the varnish or lacquer finish for some colors.

MANUFACTURER'S STATEMENT

Alpi veneers are virtually defect free. Color and grain consistency result in a high yield factor, reduced costs, and increased quality and productivity. No staining is required as the material is predyed.

Alpi Co. has been granted 3 concessions in the East of Cameroon, that equal 225,000 acres (90,000 hectares). Cameroon law obliges loggers to cut only certain species and, among those, only mature trees in designated areas. Alpi cuts only 2–4 trees per hectare, and the forest regenerates itself in only 3–4 years. While replanting programs are not enforced, a state-based organization oversees the forest, takes a census of the trees and collects taxes proportional to the surface of forest used. The funds are used for reforestation projects. 20% of the forest is available for logging.

Alpi Co. has built a purifying plant at the factory, which cleans and monitors manufacturing effluent.

PRODUCT
Synthesized alternative to exotic tropical hardwoods

MANUFACTURER
Supertech Woods, Inc.
P.O. Box 242
Schoolcraft, MI 49087
Tel (616) 323-3570

RELATED PRODUCTS
Adhesives, pp. 105–6, 119–22
Paints and Stains, pp. 136–42, 144–65
Sealers, pp. 142–3, 166–74

DESCRIPTION
EBON-X: Selected walnut hardwood, impregnated and converted into a wood product with the desirable characteristics of rare exotic hardwoods. Fully "wood-like" nature, with extreme moisture resistance and improved workability.

Color, permeated throughout, that cannot bleed or run, and can be used without protective finishes. Colors available: Ebon-X™, Ebon-X Jet Black™, Ebon-X Bronze™, Ebon-X Rose™. Material is available in dimension stock and veneer, 100% usable.

MANUFACTURER'S STATEMENT
No dyes, plastics, or solvent are used in Ebon-X manufacture. Water-borne, nontoxic chemicals are used to effect a chemical change in the cell walls.

Development of exotic hardwood alternatives will deter tropical deforestation. Continued product development will see a teak substitute, among others.

PRODUCTION

SUSTAINABLY ACQUIRED, OR RENEWABLE RESOURCE

PACKING / SHIPPING

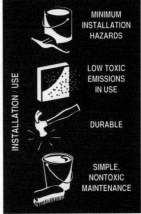

INSTALLATION / USE

MINIMUM INSTALLATION HAZARDS

LOW TOXIC EMISSIONS IN USE

DURABLE

SIMPLE, NONTOXIC MAINTENANCE

RESOURCE RECOVERY

OTHER

PRODUCTION

SUSTAINABLY ACQUIRED, OR RENEWABLE RESOURCE

PACKING / SHIPPING

INSTALLATION / USE

SEE PAGE 34

MINIMUM INSTALLATION HAZARDS

LOW TOXIC EMISSIONS IN USE

DURABLE

RESOURCE RECOVERY

OTHER

PRODUCT
Low toxicity particle board

MANUFACTURER
Rodman Industries
P.O. Box 76
Marinette, WI 54143
Tel (715) 735-9541

RELATED PRODUCTS
Adhesives, pp. 105–6, 119–22
Paints and Stains, pp. 136–42, 144–65
Sealers, pp. 142–3, 166–74

DESCRIPTION
RESINCORE I: Particle board made with phenol-based resins and sawdust. Available in 4 grades, ranging from 45–62 lbs. per cubic foot, based on ¾" thickness. Thicknesses: ⅜" to 1³⁄₁₆".

MANUFACTURER'S STATEMENT
Rodman Industries can consume 220 tons of sawdust per day, a byproduct of lumber and planing mills. Phenolic resins have no urea-formaldehyde (UF) and, since they do not emit significant quantities of formaldehyde, they are exempt from provisions of the HUD safety standard. Phenolics have a greater water and heat resistance than UF resins.

PRODUCT
Low toxicity, medium density wood fiberboard

MANUFACTURER
Medite Corporation (subsidiary of Valhi Forest Products Grp.)
P.O. Box 4040
Medford, OR 97501
Tel (503) 779-9596   Fax (503) 779-9921

RELATED PRODUCTS
Adhesives, pp. 105–6, 119–22
Paints and Stains,, pp. 136–42, 144–65
Sealers, pp. 142–3, 166–74

DESCRIPTION
MEDEX: Patented, formaldehyde-free, industrial grade MDF, manufactured with an exterior quality binder. Homogeneous wood panel manufactured from a blend of western softwoods. Water resistant for exterior use with appropriate finishing, or for interior use where low emissions are desired.

MEDITE II: Industrial quality interior MDF with no added formaldehyde. (Small amounts of naturally occurring organic formaldehyde may be present.) Homogeneous wood panel manufactured from a blend of western softwoods.

To preserve their environmentally friendly characteristics, Medite II and Medex should not be stored with products that can emit any significant amount of formaldehyde, nor should they be finished with adhesives, coatings, or laminates that can release formaldehyde.

MANUFACTURER'S STATEMENT
Medite products are resource efficient. Over 93% of the raw material used in the manufacture of Medite is derived from wood residuals from primary wood products, which were previously burned or landfilled for disposal; noncommercial wood residuals gathered incidental to a commercial harvest, which were previously burned with slash; and managed forest subcommercial-sized thinnings. No commercially viable tree has been cut solely to provide wood for Medite products.

Improved utilization of a renewable natural resource, through environmentally sound practices to produce healthy, quality, cost-effective products is Medite's ongoing goal. Medite II and Medex meet or exceed all current government regulations worldwide.

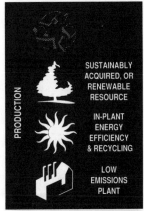

PRODUCTION

SUSTAINABLY ACQUIRED, OR RENEWABLE RESOURCE

IN-PLANT ENERGY EFFICIENCY & RECYCLING

LOW EMISSIONS PLANT

PACKING / SHIPPING

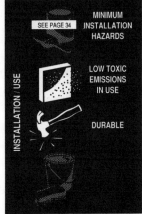

INSTALLATION / USE

SEE PAGE 34

MINIMUM INSTALLATION HAZARDS

LOW TOXIC EMISSIONS IN USE

DURABLE

RESOURCE RECOVERY

OTHER

RECYCLED CONTENT

PRODUCTION

IN-PLANT ENERGY EFFICIENCY & RECYCLING

LOW EMISSIONS PLANT

PACKING / SHIPPING

SEE PAGE 36

MINIMUM INSTALLATION HAZARDS

INSTALLATION / USE

LOW TOXIC EMISSIONS IN USE

RESOURCE RECOVERY

OTHER

RESEARCH & EDUCATION PROGRAMS

PRODUCT
Fiberboard sheathing products in varying grades
Suitable for commercial and residential uses

MANUFACTURER

Homasote Company
P.O. Box 7240
West Trenton, NJ 08628-0240
Tel (609) 883-3300
Fax (609) 530-1584

Eastern Canada:
(207) 427-6914
Western Canada:
(403) 256-4474
Ontario:
(416) 856-4994

RELATED PRODUCTS

Adhesives, pp. 105–6, 119–22

DESCRIPTION

4-WAY FLOOR DECKING: Structural subflooring where noise control is important. Multiple plies of Homasote fiberboard permanently bonded together, with tongue and groove on long edge. Standard size: 4′ × 8′ × ½″ or ⅝″.

440 CARPET BOARD, COMFORTBASE™: Carpet underlayment used to reduce impact noise and sound transmission through the floor system, applied over new or old wood subflooring. Standard size: 4′ × 8′ × ½″ or ⅝″.

N.C.F.R. HOMASOTE: Use wherever Homasote structural fiberboard can be used but where higher fire rating is desired. Standard sizes: 4′ × 8′, 4′ × 10′, 4′ × 12′. Special order sizes: 8′ × 6′, 8′ × 8′, 8′ × 10′, 8′ × 12′, 5′ × 8′, 2′ × 4′, 3′ × 4′, 4′ × 4′. Thickness: ½″.

EASY-PLY, FIRESTALL ROOF DECKING: Structural, load-bearing decking for A-frames, log homes, and room additions with exposed beam construction. Multiple plies of high density Homasote structural fiberboard with 6 mil prefinished, white vinyl film surface. Face ply of N.C.F.R. Homasote. Standard size: 2′ × 8′. Thicknesses: 1″, 1⅜″, 1⅞″, 2¹⁄₁₆″.

MANUFACTURER'S STATEMENT

Homasote is made from 100% recycled newspaper, with no formaldehyde or asbestos additives. Heavy metal content is less than 100 parts per million. It is fully biodegradable, except for products laminated with other materials.

Our manufacturing process dates back to 1909. Homasote building products help in the conservation of more that 1,370,000 trees per year, and eliminate more than 160,000,000 lbs. of solid waste annually.

Homasote products are 100% environmentally safe, contributing no contaminants or pollutants to the atmosphere. They are energy saving, nailable, and easy to install. Water removed from products during manufacturing is recycled. Processing equipment is state of the art: efficient and technically advanced. Constant product innovation is sustained through ongoing research and development.

PRODUCT
Mixed waste paper and vegetable fiberboard sheathing panels

MANUFACTURER
PanTERRE America Inc.
2700 Wilson Blvd.
Arlington, VA 22201
Tel (703) 247-3140   Fax (703) 247-3178

RELATED PRODUCTS
Adhesives, pp. 105–6, 119–22
Paints, pp. 136–42, 144–60
Sealers, pp. 142–3, 166–74

DESCRIPTION
PANTERRE: Similar to drywall panel, but with superior acoustical and thermal properties, this vegetable fiber panel is suitable for a variety of building purposes, such as underlayment and sheathing, as well as furniture uses. It can be finished with various materials, such as plastic laminates, cardboard, and wood veneers.

The core of this construction panel consists of post-consumer paper waste and ground vegetable fiber waste from agricultural industries, such as rice husks, peanut shells, or wheat straw. The cellulose mixture is hydropulped and formed into panels. Materials are heated and compressed, and no adhesive bonding agents are used. Natural fungicides, fire retardants, and rot preservatives are added, if needed, to meet a specification. Faces are of kraft paper bonded to the core with organic glue.

Board dimensions are standard 4' × 8', and thicknesses available are ½" and ¾". Fiberboard has sound deadening properties, as well as an R-value of 2 per inch. A fire retardant additive, boric acid, gives the panel a Class B fire rating.

MANUFACTURER'S STATEMENT
The operation consumes up to 14.8 tons of waste paper per day (4,200 tons per year) and approximately the same amount of vegetable material. Process water is continuously treated and recycled into production so that there is no effluent discharge. A heat exchanger in the kiln captures the steam used to dry the boards, and is used to heat the plant in winter. Dust from the sanding process is recovered with a vacuum, and is sent back to the hydropulper.

PanTERRE is 100% recyclable, and chemical stability is guaranteed without any toxic binders, asbestos, or formaldehyde.

PanTERRE is presently imported from Belgium, and PanTERRE America Inc. expects to license a North American production facility in 18–24 months. The European parent company, Terre, is a 45-year-old cooperative that has created self-sustaining economic development projects in Belgium and in developing countries. Terre's research and development yielded PanTERRE, a value-added, simple, low-tech product which could be made from collected recyclables, using a high proportion of unskilled and semi-skilled labor. The product has won many innovation awards, and is certified with the Blue Angel seal in Germany as an environmentally sound product.

PRODUCTION

RECYCLED CONTENT

IN-PLANT ENERGY EFFICIENCY & RECYCLING

LOW EMISSIONS PLANT

PACKING / SHIPPING

SEE PAGE 36

MINIMUM INSTALLATION HAZARDS

LOW TOXIC EMISSIONS IN USE

INSTALLATION / USE

RESOURCE RECOVERY

OTHER

FAIR BUSINESS PRACTICES

INTERIOR PANELS
Mixed Fiberboard

RECYCLED
CONTENT

PRODUCTION

LOW
EMISSIONS
PLANT

PACKING / SHIPPING

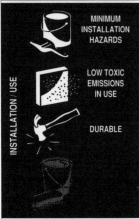

MINIMUM
INSTALLATION
HAZARDS

LOW TOXIC
EMISSIONS
IN USE

DURABLE

INSTALLATION / USE

RESOURCE RECOVERY

OTHER

PRODUCT
Decorative architectural materials of cellulose and mineral fillers (brackets, moldings, columns, planters)

MANUFACTURER
FiberStone Quarries, Inc.
P.O. Box 1026, 1112 W. King St.
Quincy, FL 32351
Tel (904) 627-1083, (800) 621-0565   Fax (904) 627-2640

DESCRIPTION
QUINSTONE: Cast stone consisting of recycled mineral or fiber fillers (depending on availability) combined with water-based binders and non-toxic, mineral color pigments. Available in 4 textures, and standard or custom colors.

Precast, presealed architectural components include matched veneer wall panels, molding, accessories, etc. Custom variations, sizes, and items available. Quinstone is lightweight, and fire-rated as zero flame spread. Installation is easily done with carpentry tools.

Recycled contents include calcium carbonate, fiberized post-consumer newspaper, lint (waste byproduct of cotton milling), or cellulose (waste byproduct of paper-making).

MANUFACTURER'S STATEMENT
Quinstone was originally developed as a display system for glass art objects. The concept was to use materials that would reduce solid waste from other industries, as well as keep processes and admixtures nontoxic for a "clean" product, and healthier working conditions.

PRODUCT
Decorative architectural elements of cellulose and mineral fillers

MANUFACTURER
Syndesis, Inc.
2908 Colorado Ave.
Santa Monica, CA 90404
Tel (213) 829-9932

RELATED PRODUCTS

SYNSORY is a line of accessories (bowls, vases, soap dishes, clocks, candle holders, picture frames, clocks, children's toys, and office products).

STANDARDS will be available in 1992, a new line of lower priced, molded tabletops and planters.

DESCRIPTION

SYNDECRETE SOLID SURFACING MATERIAL: A lightweight, cement-based composite using natural minerals and incorporating up to 30% recycled materials. Recycled filler materials can include post-consumer and scrap materials from industry, such as plastic regrinds, wood chips, crushed glass, metal shavings, or stone fragments. Reinforced with a polypropylene fiber waste product from carpet manufacturing. Fly ash, a residue from burning coal to produce electricity, is added, reducing cement consumption by 20%. Contains no resins or polymers.

More resistant to chipping or breakage than conventional concrete, tile, or limestone and can be worked with woodworking tools. Suitable for interior and exterior use. It has excellent thermal and acoustic properties, and is heat and fire retardant. Sealed with an FDA approved water-based acrylic sealer, making the surface resistant to water penetration, absorption, and staining. Available in a wide variety of colors and finishes.

Products include custom surfaces, moldings, surrounds, tiles and wall panels, counter tops, table tops and bases, planters and landscape elements, furniture and fixtures, bathtubs and sinks, fireplaces, fountains, and stairs.

MANUFACTURER'S STATEMENT

Syndesis, Inc.'s corporate values are based on respect for and efficient use of human and natural resources. Environmental awareness permeates every aspect of its approach to business, from its products and manufacturing processes to marketing and promotional materials, and architectural designs. Syndesis is committed to transforming the excesses and waste of today's society into value-added, quality designed products.

Syndecrete was developed as an alternative to limited or nonrenewable natural materials such as wood and stone, and synthetic, petroleum-based solid and laminating materials. It is chemically inert and not subject to off-gassing.

Recycling efforts include the filtering and recycling of polishing water, recovery and reuse of form building materials, and the collection and reactivation of dust. Packaging materials are generated by shredding in-house waste paper. All printed promotional materials are printed on recycled paper using soy-based inks.

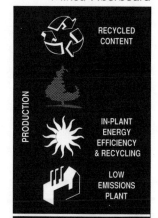

PRODUCTION

RECYCLED CONTENT

IN-PLANT ENERGY EFFICIENCY & RECYCLING

LOW EMISSIONS PLANT

PACKING / SHIPPING

MINIMUM, RECYCLED, RECYCLABLE PACKAGING

INSTALLATION / USE

MINIMUM INSTALLATION HAZARDS

LOW TOXIC EMISSIONS IN USE

DURABLE

SIMPLE, NONTOXIC MAINTENANCE

RESOURCE RECOVERY

OTHER

FAIR BUSINESS PRACTICES

RECYCLED
CONTENT

PRODUCTION

IN-PLANT
ENERGY
EFFICIENCY
& RECYCLING

LOW
EMISSIONS
PLANT

PACKING / SHIPPING

SEE PAGE 37

MINIMUM
INSTALLATION
HAZARDS

INSTALLATION / USE

LOW TOXIC
EMISSIONS
IN USE

DURABLE

RESOURCE RECOVERY

OTHER

PRODUCT
Cellulose-reinforced gypsum wallboard (drywall)

MANUFACTURER
Louisiana Pacific
111 S.W. 5th Ave., Suite 4200
Portland, OR 97204
Tel (503) 221-0800   Fax (503) 796-0204

RELATED PRODUCTS
FIBERBOND DRI MIX JOINT COMPOUND: Two-step joint system involves applying the compound over joint, followed by a single application of a reputable ready-mix compound. The result is a uniform surface. No joint taping is required, although joints may also be finished using traditional methods.

DESCRIPTION
FIBERBOND: Fiber-reinforced gypsum panel for use in interior walls and ceilings, in standard and fire-rated applications.

80% of the fiber is from nonde-inked post-consumer newsprint and 20% from nonnewsprint paper. Paper is fiberized by running it through a hammermill refiner twice. Currently natural gypsum is used. However, testing with German technology is under way, experimenting with flue-gas gypsum recovered from coal-burning electricity plants. Starch and other natural, proprietary additives are incorporated.

Can be installed with nails, screws, staples, or adhesives over wood, steel, or masonry substrates. Available in standard sizes: 4' × 8', 4' × 9', 4' × 10', and 4' × 12'. Thicknesses: ⅜", ½", and ⅝" type X as per ASTM C36. Product is a 3-layer board: outer layers contain only cellulose and gypsum, while middle layer contains perlite to lighten the board. Overall board density is 53–55 lbs. per cubic foot.

MANUFACTURER'S STATEMENT
Louisiana Pacific is working to produce innovative, affordable products that are environmentally sound. Their goal is to provide products that are better, more cost effective, and easier to install than conventional materials. The company is very concerned with the rising costs of house building and is actively promoting awareness and encouragement of affordable housing.

Little waste water is produced with the semi-dry process, which involves calcining the gypsum and rapidly combining it with the dry cellulose. Curing begins to occur immediately. With this process, no sludge or slurry is produced to require disposal.

(See also INSULATION, p. 29.)

CARPETING

Carpeted floors have been associated with comfort and luxury for hundreds of years. The appearance, feel, warmth, and sound-absorbing properties of a textile underfoot have obvious value, but come with a price in terms of environmental, maintenance, and health factors.

Traditional carpets were woven on looms with plant and animal fibers. They were made in limited sizes, depending on the size of the loom, and were placed on top of a finished floor of wood, tile, or stone. The materials used were renewable, biodegradable and, as long as they were kept dry and clean, the carpets had little impact on health. They would be kept clean by taking them outside for a seasonal "beating" to remove dust, or for cleaning with a brush and soapy water. High quality carpets handled in this way could last for several generations.

Today most carpets are made of synthetic fibers from petroleum sources. They are made on continuous looms, usually by bonding a fiber to a backing with flexible latex glue. The materials are neither renewable nor biodegradable, and gases emitted from new materials are a serious cause of indoor air pollution. In most uses today, the carpet and undercushion are fixed to an unfinished wood or concrete subfloor with glue or tacks, and are never removed until they are discarded. The cleaning processes for fixed carpets range from aerosol "spot cleaners" containing hazardous solvents to sophisticated steam, water, or electrostatic machinery using chemical soil removers, brighteners, perfumes, antibacterial agents, and stain-resistance treatments. When synthetic carpet is discarded, usually after four to twelve years, it is dumped in landfills, where it will remain intact for several generations.

Carpeted floors always require more maintenance than other floor types, and have an inherent ability to trap dust and soil. Under moist conditions, the soil will support the growth of bacteria and fungi (such as mildew), which leads to a chronic "sour" odor and the release of highly allergenic fungal spores. Even under dry conditions, carpet which is not cleaned regularly will become a major source of dust and odors. Due to their porous structure and large surface area, carpets also trap odors and re-release them very effectively. This is one of the reasons, for example, that carpet which has been exposed to cigarette smoke is so difficult to clean.

Recent developments in carpets could begin to rectify some of their inherent health and environmental problems. Quality area carpets with bound edges, placed on finished wood or tile floors, are beginning to return as an important element of interior design. This type of installation helps to reduce in-home cleaning, and encourages cleaning services which remove the carpet for cleaning under well-controlled conditions, with the least hazardous cleaners.

A return to natural materials like wool, sisal, cotton, and grasses is also apparent in the floor covering industry. These materials are renewable, typically lower in toxic content than their synthetic counterparts, last very well if they are of high quality, and are biodegradable when finally discarded.

In the synthetic fiber industry, new advances include carpet made from recycled plastic, several types of lower toxicity carpets made without the latex bonding, removable fastening systems which can extend carpet life (see INSTALLATION METHODS, p. 56), and serious efforts to produce a nylon carpet construction which is recyclable.

FIBER TYPES

NATURAL FIBERS

Natural fibers are from renewable sources, use little energy and produce little pollution in processing, and are biodegradable when discarded. One disadvantage, however, is that some are not very durable, and some may produce allergenic dust.

WOOL

Raising sheep for wool creates environmental problems in some regions, such as land deforestation for grazing and the ensuing erosion from loss of tree coverage. This is particularly serious on steep terrain. On suitable pasture ground, however, sheep have very little environmental impact if the land is not overgrazed. The major problems arise with pesticide sprays and dips used to treat the sheep.

Most wool used in high quality carpets is produced by shearing live animals annually. Wool from slaughtered animals is of lower quality, and is sometimes removed from the skin using chemical treatments or bacterial applications. This wool is blended with sheared fiber.

Wool is refined for use in carpet by "scouring" with detergents or solvents which remove most of the natural oil (lanolin), "carbonizing" with sulfuric acid to remove tough dirt, a final washing and drying, and carding and combing in preparation for spinning. These processes produce waste water that must be treated. After spinning, wool carpet is produced either by pressing the yarn into a mesh backing and bonding it in place with latex, or by more expensive, traditional knotting methods. (See CARPET CONSTRUCTION, p. 53, and BACKING, p. 54.)

Wool fiber is very durable, and is naturally fire resistant. These qualities simplify the treatments needed for it to meet performance requirements and fire code restrictions. Wool is, however, prone to attack by moths and is often mothproofed before sale. Though mothproofing is done with toxic materials, such as sodium fluorosilicate, these are not as much of a health risk in carpets as they are in fabrics which contact the skin. (See FURNITURE, p. 211.)

COTTON

Cotton is both a very heavy "feeder" (that is, it depletes soils), and prone to attack by many insects, particularly weevils. Commercial production therefore depends heavily on chemical fertilizers and pesticides. Chemical defoliants are often used at the harvest stage to make the crop easier to pick. One area for possible improvement is through breeding programs and returning to traditional varieties which, though they may produce less cotton, are more drought resistant and require less fertilizer and pesticide.

Cotton accepts dye easily and is also subjected to a wide range of treatments for flame retardancy, stain resistance, wrinkle resistance, and sizing, most of which cause further environmental damage and add to exposure risk for production workers and users. Though cotton has traditionally been used only for handcrafted floor coverings, it is now possible to buy commercial cotton carpets.

OTHER PLANT FIBERS

Plant fibers are either collected from their wild state or raised on plantations with varying degrees of harvesting and replanting management, and erosion control. The fibers can be used untreated, though they are often dyed or treated for insect repellency or stain resistance.

Though plant fiber carpets are usually associated with rustic interiors, there are several sisal carpet manufacturers who make fine-textured, very high quality and very durable products. These are appropriate even with a very refined interior decor.

The major problem with plant fibers is their lack of stain resistance. The fiber is very porous and absorbs liquid spills. It may also fade or darken in sunlight. Spray-on sealers are available to reduce the porosity of the fiber, but they are made of toxic solvents. The safest approach is to use plant fiber carpet in areas where food spills are unlikely to occur. Due to their open weave construction, most plant fiber carpets are actually easy to maintain, as soil tends to fall through it, and is easily vacuumed. All plant fibers are biodegradable, so worn-out carpet can be simply chopped up and composted.

SISAL: This is a tough fiber from the leaves of the *Agave* plant, grown in the West Indies and elsewhere. Sisal is used to produce very durable floor matting. Traditionally used as nonslip carpeting for stair runners and hallways, it is now available in a range of colors and textures as area carpet. Good quality sisal carpet in heavy traffic areas can maintain its appearance and outlast cut-pile and loop carpets by years. Sisal has good potential as a sustainably managed crop and for wider use in floor coverings.

COIR: The fiber from coconut husks, coir is another very tough material, and it is a waste product of coconut harvesting. Coir is generally used for lower quality floor coverings, such as rustic matting and door mats, because it is somewhat coarse. Shredded coir is also used for filling and as packing material.

HEMP: This fiber from the *Cannabis* shrub, grown in Asia and elsewhere, forms a strong cordage which is used to make rope mats, and for backing and reinforcing carpets and mats. Hemp is grown for fiber in many climates, but the fiber quality does not lend itself to as many floor covering applications because it is so coarse.

JUTE: The fiber from the *Corchorus* shrub, a member of the Linden family which grows in Asia and the South Pacific, jute is the most common fiber used for carpet secondary backings.

REEDS AND GRASSES: A wide range of reeds and grasses are also used for woven matting. These products are usually inexpensive, but they are not very durable. Reeds and grasses are harvested from the wild and have good potential for sustainability.

SYNTHETIC FIBERS

Though some synthetic textile fibers are made from wood fiber (the cellulosics), nearly all common commercial synthetic carpet fibers are made from oil. (See FURNITURE, p. 212, for more information on synthetic fibers.) The use of nonrenewable petroleum depletes a precious resource and necessitates dependence on imports. The extraction of oil, its processing, and final synthesis into fibers like nylon and polyester also produce serious air and water pollution, as well as solid waste. The people who work in these industries are exposed to carcinogenic materials, heavy metals, and dangerous acids and solvents. (See TREATMENTS, below.)

NYLON

Nylon is a large family of polyamide resins, of which only two have major commercial significance. Nylon 66 accounts for 75% of nylon production; most of the remaining 25% is nylon 6. Nylon can be formed into monofilament thread, or drawn into fine fibers and spun into yarn.

Several different types of carpet construction use nylon as a cut-pile or looped "face fiber." Continuous filament nylon (CFN) and bulked continuous filament (BCF) refer to fibers that are continuous and locked into the carpet backing, creating a clean, pill-free surface that will not fuzz or shed. Nylon is the most widely used fiber for quality synthetic carpets due to its durability.

POLYESTER

Polyesters are produced by reactions between acids and alcohols, derived from petroleum. The environmental impact is similar to that of other synthetic fibers. However, most polyester fiber carpet is less durable than nylon and begins to lose its new appearance quickly in traffic areas. For this reason it should be used only in low traffic areas. Polyester carpets made from recycled materials are now available in modestly priced lines. These are made from remelted soft drink bottles, and are comparable in appearance, quality, and durability to those derived directly from petroleum.

POLYPROPYLENE

This fiber is produced very cheaply by directly polymerizing (linking into long molecule chains) propylene derived from distilled oil. Because it is a simple process, it has somewhat lower environmental impact than other synthetic fiber production, though dyeing and treatment are similar. Polypropylene fiber carpet is of lower cost, but is usually more abrasive than polyester or nylon because it is a very hard fiber. It is therefore not often used where a soft surface feel is desired. It also begins to pill and shed sooner than good quality nylon.

MIXED FIBERS

Synthetic and natural fibers are sometimes blended. Nylon/polyester, nylon/polypropylene, wool/nylon and polypropylene/wool are common blends which increase durability and reduce cost. Unfortunately, emerging efforts to deal with carpet waste by recycling the face fiber are made more difficult by this practice. At the very least, the value of the recycled material would be degraded by fiber mixes. Wool/synthetic fiber mixes are not practical to recycle at all. The value of these products is that they are expected to outlast most single-material fibers, both in terms of appearance retention and durability.

CARPET CONSTRUCTION

Prior to the 1960s, virtually all carpets were woven or knitted with natural fibers. However, these methods and materials could not meet increasing consumer demands for low cost and high performance. The high cost of wool production and the low efficiency of traditional weaving methods are severe limitations to producing reasonably priced carpet.

The tufted loom method was developed because it has ten times the production efficiency of weaving. In this process, synthetic pile fibers (nylon, polyester, and polypropylene) are stitched through a jute or polypropylene primary backing, sealed with a latex bond (either natural or synthetic), and covered by a secondary backing or protective scrim of jute or polypropylene. The majority of carpet is now made by this process, but there are recent innovations. One of these is the introduction of a method of heat-welding the fiber to a synthetic or foam cushion backing. (See BACKING, below.)

Handmade carpets and factory production using traditional materials and weaving methods are returning to the market in greater quantity recently. Though these are expensive, the demand for high quality, durable, classic construction is increasing along with consumer awareness of the negative environmental impacts caused by the production of synthetic fibers.

TREATMENTS

A large number of treatments are applied to carpet fiber during production, such as bleaches, dyes, mothproofing (wool only), stain resistants, and antistatics. All these steps add not only to the chemical exposure of plant workers, but dramatically increase the volume of plant waste water to be treated.

Dyeing is the major treatment that a carpet undergoes, and methods vary greatly. It can occur at three different stages of production, and the chosen method determines the amount of waste water produced. Solution dyeing involves adding pigment to the liquid synthetic before it is spun. This is not an extra step in the process, and results in virtually no added waste. There are several types of stock dyeing (dyeing the actual yarn). Yarn dyeing uses three to four times as much processing water per pound of floor covering material than continuous dyeing, where very large quantities are colored at once. With piece dyeing, a manufactured carpet is immersed in a dye bath, inefficiently utilizing the dyestuff and leaving large quantities of waste water.

Heat-setting of fibers, to help them maintain their shape, is another process which uses large amounts of water for cooling. Stain-resistant finishes applied to the fiber surface in the plant also require an additional treatment bath and after-rinse.

Early formulations for stain-resistance treatments used hazardous polymers and glycols as solvents. In the past few years, safer chemistry has been developed to reduce toxic exposure. The quantity of water required has not decreased, however, and water recycling and treatment is one of the main environmental challenges of carpet production. Some of the measures currently being developed are the reuse of dye and finish baths to reduce consumption, and electrochemical coagulation/filtration of waste water, or biological treatment systems to improve effluent quality. Improved dye and stain-resistance methods which are part of the fiber instead of topical are less wasteful and reduce effluent problems.

Finished carpets are treated to control insect attack and bacterial contamination during use. These treatment chemicals are not only a potential health hazard, but they have the added consequence of encouraging more resistant organisms. An insecticide or repellent treatment does not kill or repel all insects. There are always a few in every population which can tolerate the poison, and these are the ones which survive and reproduce. Through successive generations, due to the genes inherited from the parents, the offspring are even more likely to be resistant to the chemical which was intended to kill or repel them, until finally the treatment has little effect on the majority of the population. This pattern is true for antibacterial treatments as well, and has been thoroughly proven in hospitals, where a great deal of antibacterial solutions is applied to walls and floors.

Treatments are not a substitute for wise choices in the use of interior finish materials. For example, floor and wall coverings which retain moisture are not appropriate in any use where there is frequently excess moisture present.

FLAMMABILITY

Flammable fibers used in carpet must be treated with fire retardant, by law. Glycols, phosphorous, chlorine, and bromine-based compounds are commonly used for this purpose, and some of these substances are very hazardous to produce and handle. They are usually well fixed in the material, but can be a source of indoor air pollution.

BACKING, UNDERLAYMENT, & UNDERCUSHION

BACKING

The latex bonding used in most carpet backing today makes up a significant part of the indoor air pollution produced by new carpet. Both natural latex from tropical rubber trees and synthetic latex called styrene-butadiene rubber (SBR) are used. Natural latex has a strong odor in its raw form due to naturally occurring volatile substances in the rubber tree's sap. A significant number of people are allergic to natural latex odors and skin contact. Most of the odor of natural latex can be removed by refining, however, and manufacturers of quality carpets use only well-refined material.

Synthetic latex (SBR) is a source of several volatile agents which contribute to the characteristic "new carpet" odor. Of about 100 different gases which have been identified from new carpet made with synthetic latex, one of the most prevalent is called 4-phenylcyclohexene (4-PC). This gas is an accidental byproduct of synthetic latex manufacturing and, according to latex manufacturers, could be nearly eliminated by changes in the production process. Though these changes would result in only a small cost increase to carpet production, they have not been implemented by any manufacturers to date.

An important advance in synthetic carpet production is a technique called "fusion" construction or "reaction bonding." In this process, the heat-sensitive face fiber is softened by heat as it is pressed into a synthetic or foam cushion backing. The fiber and backing then become heat-welded together, reducing the amount of material in the carpet and possibly the need for undercushion. It also eliminates latex, which is a source of pollution during manufacture, produces emissions during use, and is an obstacle to recycling.

UNDERLAYMENT

Composition wood products are often used as underlayment over rough subfloors before laying carpet, to improve the surface and cover cracks. Standard particle boards are used for this purpose, even though they are known to be the primary source of formaldehyde gas in new homes. Formaldehyde is emitted by the glue used in making the board, and is released into the room air because it is essentially unrestrained by porous carpet covering. Exterior grade plywood or formaldehyde-free wood fiberboard are superior choices for underlayment. Wood fiberboards are available with recycled contents. (See INTERIOR CONSTRUCTION PANELS, p. 35.)

UNDERCUSHION

The purpose of undercushion is to improve the resilience of the carpet, enhance its sound-absorbing ability, preserve its appearance, prolong its life, and allow easier removal and reuse. Most undercushions are made of foamed plastic or rubber. These contain petroleum products which cause serious air and water pollution at all stages of their acquisition and production, as well as the depletion of a limited resource. Two common foamed plastics used for undercushion are polyurethane (isocyanurate) and polyvinyl chloride (PVC). Latex rubbers (natural and synthetic) and other rubbers are also foamed for heavier duty commercial grade cushion.

Plastics and rubber are foamed with gases called "blowing agents." These can be chlorofluorocarbons (CFCs), hydrocarbons like pentane and butane, safe, inert gases like nitrogen and carbon dioxide, or simply steam or air. Unlike upholstery grade foams and plastic building insulations, undercushion foams have traditionally not been made with ozone-depleting CFCs.

Another family of undercushions are those made with fiber compressed into a felt. These can be made from polyester fiber, natural fiber like hemp, or mixed fiber from clean carpet mill waste. These undercushions are available in residential and commercial grades and perform quite well.

The major environmental concerns with undercushion are the source of the material and the disposal problems it may present. Important advantages of the felt undercushions are that they use less material, and can be made from mill waste. Recycling is not currently an option for foamed plastic and rubber materials, but they could theoretically be chopped up and reused as loose fill insulation, or compressed into industrial rubber mats. At least one manufacturer is now making a carpet undercushion from the chipped rubber of discarded tires. It has a distinct sulfur-like odor when fresh, but this declines rapidly.

Health risks from undercushion are generally far less than from the carpet itself, though some heavy, commercial rubber undercushions are made from synthetic latex (SBR) and produce the same sort of emissions as the carpet backing. (See BACKING, above.) Generally the felt undercushions are more chemically stable and produce less air pollution than the foamed types.

Some carpet lines are now available with a nonporous plastic foam cushion, which is formed as part of the carpet itself. This construction can be found in commercial quality carpet and is made from a low emission polyvinyl chloride (PVC) material with a self-adhesive backing. There are important advantages to this type of construction because it does not require adhesive, and it is removable without destroying the carpet or the substrate.

Undercushion installation is otherwise similar to carpet installation (see below). If the undercushion is glued down it is not recoverable for reuse. A durable cushion, such as fiber felt, commercial rubber, or recycled rubber will last long enough to be reused if it is looselaid.

CARPET EMISSIONS

Volatile emissions from carpets, like those from other interior materials, do diminish with time. One measure of emissions is called a "half-life," meaning the time required for an emission to decrease by half. Because some of the emissions from new carpet may have a half-life of a few days or weeks, emissions can be reduced by airing out the carpet before installation. This is usually done by rolling out the material in an area with good ventilation. Some carpet suppliers will do this by special arrangement with the purchaser. The quantity of gases released after installation can be reduced by this method, but the amount depends on ventilation, temperature, and humidity.

In the final analysis, there is no real substitute for the reduction of toxic emissions at their source. For carpet production and installation, this means low emission carpet construction, and low toxicity installation and cleaning methods.

INSTALLATION METHODS

In many commercial installations and in some residential uses, carpet is laid semipermanently by gluing it with a wet adhesive to a wood or concrete surface. The adhesives are either solvent-based types which are hazardous to handle, or safer water-based latex types. In some cases, a double glue-down method is used, where the undercushion is first glued to the floor and then the carpet glued to the undercushion.

Opportunities for the reuse of carpet are lost by direct glue-down installation because the carpet is unusable when removed. This process uses very large amounts of glue and exposes tradespeople to high levels of volatile chemicals, some of which are hazardous. The emissions from carpet adhesives generally continue for several weeks to months, thereby also exposing the building occupants.

Two important advances in installation methods have now made wet adhesive methods unnecessary in many cases, and both allow the removal of carpet for reuse. One method is a "dry adhesive" backing, which is applied to a cushion-backed carpet during manufacture. The adhesive is protected until application by a thin sheet of polyethylene which is removed before installation. This carpet type is available in tiles (precut squares of carpet which are laid like tile), and in rolls. Another method uses a hook and loop system similar to Velcro, in which the loop half is part of the carpet backing and the hook half is glued to the floor in thin strips. A small amount of this type of fastening, around the edge and at intervals, will hold a large area of carpet in place. Some commercial building managers using these systems rotate carpet from low wear areas to high traffic areas to extend carpet life.

Where conventional glue-down methods are still desired or necessary (and for other types of glue-down floor coverings), there are carefully formulated, low toxicity adhesives available which can substantially reduce the exposure to installers and building occupants. (See INSTALLATION MATERIALS, pp. 106, 116–22.)

Conventional "nail strip" installations have important advantages over glue-down methods because they are removable and avoid the use of glue. The major problem with nail strip fastening is that the traditional stretching procedure, using a kneepunch stretcher, has short- and long-term physical effects on workers. Carpet layers have well-recognized patterns of knee, back, and muscle injury as a direct result of carpet stretching. Less labor-intensive installation methods like the self-adhesive ones are therefore better for tradespeople. Stretching machines are available, which can prevent injuries to carpet layers, but few installation companies invest in them.

MAINTENANCE

Wet or steam cleaning of carpet requires the use of many types of cleaning chemicals, and produces substantial amounts of contaminated waste water. Post-cleaning treatments usually include stain-resistance agents (discussed in TREATMENTS, p. 53) and perfumes. One of the principal problems with wet cleaning methods is the control of moisture. There must be enough moisture to clean the carpet effectively, but not too much or it may damage the carpet and other materials. Residual water is also a risk because, if it takes too long to dry, it may become sour or support fungus.

Dry solvent cleaning systems are also used for both spot cleaning and general cleaning. These use a solvent which turns to a powder when dry, and can then be vacuumed. This method is not adequate for general cleaning because it does not fully remove soil from the fiber. The solvents introduce handling and safe disposal risks.

RESOURCE RECOVERY

Approximately 70% of carpet manufactured today is used to replace existing carpet. The average life for all carpets is eight years. This cycle is determined in part by the service life of the carpet, but is strongly influenced by construction and real estate activity, current fashion and innovation, and economic conditions. In large urban areas, carpet represents about 3% of the total volume of landfill waste. Carpet buried in a landfill may last for hundreds of years.

RECYCLABILITY AND CONVERSION

Recycling carpet may become feasible when methods of separating the face fibers from the backing are developed. With current technology, nylon face fibers would be so weakened by cleaning, bleaching, and remelting them into pellets for forming new fiber that they would not be useful for carpet. A stronger molecular nylon configuration would be required to make recycling possible, and this is now being researched. Carpets with polypropylene face fibers have more potential for recycling, particularly those with polypropylene backings. If they are bonded with latex, separation remains a problem. This is another reason the new fusion welding method of bonding fiber to backing is important.

There are existing opportunities for the conversion of unused textile waste. Polypropylene selvedges have characteristics similar to polystyrene and can be made into some lower grade plastics, such as cassette tape covers, cord, and rope, carpet backing, and containers. Clean jute waste can be ground and composted, as it is fully biodegradable. Clean wool waste is easily recycled into lesser quality wool textile products, such as blended fiber fabrics, although the quality is not high enough for carpet. Clean nylon scrap can be recycled into other commercial nylon products, although, again, it is not of high enough quality for carpet. The conversion of clean carpet scraps of all kinds into stuffing and packing materials has been a common practice for some time.

Other waste plastic materials are currently being converted into carpet fiber, on a small scale. One plastics recycler in the U.S. is reclaiming polyethylene terephthalate (PET) soft drink bottles and forming them into pellets, which some carpet manufacturers are using to produce a polyester carpet of equal quality to other polyesters.

PRODUCTION AND INDUSTRY IMPROVEMENTS

Carpet mills in North America are concentrated in a few areas, mostly in Georgia and California. Water utilization concerns, and effluent and solid waste disposal problems, have led to necessary innovations. Existing landfill restrictions, for example, and the high costs associated with preventive measures for new landfills have forced attention on the reuse and minimization of mill scraps, packaging, and other solid wastes.

Among the solutions are higher loom standards to produce more usable material, conversion of waste to useful fiber such as underpad and packing materials, and using scrap as fuel for curing ovens and to produce steam for the production facility. The carpet industry acknowledges that burning waste is a low value use for scrap material. Fiber separation technology is being explored, which would allow waste to be reused for higher quality purposes.

WORLD MARKET

Producing the raw material for natural plant fiber carpets like jute is relatively labor intensive, and is generally done in poorer countries where working conditions and wages are also poor. There are, however, a few cooperative companies emerging which ensure better treatment of laborers and even a share of ownership.

The lack of internationally accepted environmental and labor standards is another serious obstacle to fair world trade. Goods from countries which do not have adequate standards can be imported without tariff penalties. This results in unfair competition for goods manufactured in the U.S., Europe, and Canada, where emissions controls, higher wages, and a better standard of living for employees add to production costs. Countries with minimal standards, like Thailand and China, increase production as they earn a larger market share, leading to higher pollution levels. Though these countries need dollars from abroad, the money does not really help the people who need it most. They are exposed to more hazards, while their actual income declines in terms of ability to support a family.

SUSTAINABLY ACQUIRED, OR RENEWABLE RESOURCE

PRODUCTION

MINIMUM, RECYCLED, RECYCLABLE PACKAGING

PACKING / SHIPPING

SEE PAGE 56

MINIMUM INSTALLATION HAZARDS

INSTALLATION / USE

RESOURCE RECOVERY

OTHER

PRODUCT
Residential quality wool broadloom carpeting

MANUFACTURER
Cavalier Bremworth Ltd. (New Zealand)
Imported by
Bremworth Carpets
1776 Arnold Industrial Way
Concord, CA 94520
Tel (510) 798-7242

Colin Campbell & Sons
202-1717 W. 5th Ave.
Vancouver, BC V6J 1P1
Tel (604) 734-2758
Fax (604) 734-1512

RELATED PRODUCTS
Adhesives, pp. 105–6, 116–23
Undercushion, pp. 55, 76–78

DESCRIPTION
CASTLEMOOR, SHEREZADE: 100% pure New Zealand wool, tweed hard-twist construction in 12′ (366cm) wide goods, with jute backing and synthetic latex finishing. Recommended for heavy duty residential use. Available in 9 color combinations.

CUMBERLAND, COLONIAL HOUSE: 100% pure New Zealand wool, plain hard-twist construction in 12′ (366cm) wide goods, with jute backing and synthetic latex finishing. Recommended for heavy duty residential use. Available in 16 solid colors.

KINLOCK: 100% pure New Zealand wool, tweed high-low loop pile construction in 12′ (366cm) wide goods, with jute backing and synthetic latex finishing. Recommended for heavy duty residential use. Available in 8 two-tone (natural and light color) combinations.

KARATEX: 100% pure New Zealand wool, level loop pile construction in 12′ (366cm) wide goods, with jute backing and synthetic latex finishing. Recommended for heavy duty residential use. Available in 8 natural and pastel heathered colors.

MANUFACTURER'S STATEMENT
Packaging consists of polyethylene film (recyclable) and hessian, which can be reclaimed and converted into felt underlay or will biodegrade in landfills.

The product is designed for long life. It can also be reclaimed for conversion to felt underlay. Bremworth stands behind the policy of fair, honest business practices at the manufacturing stage as well as international trading levels.

The Drysden sheep, a breed of sheep developed in New Zealand, produces a wool fiber which is hollow, making it more resilient and flexible.

PRODUCT
Wool broadloom carpet

MANUFACTURER
Gaskell Carpets Ltd. (UK)
Imported by Colin Campbell & Sons
202-1717 W. 5th Ave.
Vancouver, BC V6J 1P1
Tel (604) 734-2758

RELATED PRODUCTS
Adhesives, pp. 105–6, 116–23
Undercushion, pp. 55, 76–78

DESCRIPTION
HERDWICK: 100% natural wool level loop (Berber style construction) face yarn is undyed and unbleached. Undyed goods are available in 3 subtle natural shades of gray. Backing is natural latex with synthetic drying compounds.

MANUFACTURER'S STATEMENT
Lambs are born with a black fleece, but with maturity their wool turns to varying shades of gray. When blended during manufacturing, these natural shades produce color effects without the use of dyes.

Owner of a herd of Herdwick sheep, renowned author of children's books Beatrix Potter left her farm to the National Trust (a British organization concerned with conservation since 1895). The National Trust protects nearly a third of the land in the Lake District of Nr. Sawrey and leases to its tenant farmers one third of all Herdwick sheep—some 25,000 head.

PRODUCTION

SUSTAINABLY ACQUIRED, OR RENEWABLE RESOURCE

LOW EMISSIONS PLANT

PACKING / SHIPPING

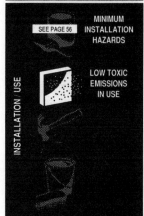

INSTALLATION USE

SEE PAGE 56

MINIMUM INSTALLATION HAZARDS

LOW TOXIC EMISSIONS IN USE

RESOURCE RECOVERY

OTHER

PRODUCTION

SUSTAINABLY
ACQUIRED, OR
RENEWABLE
RESOURCE

PACKING / SHIPPING

INSTALLATION / USE

SEE PAGE 56

MINIMUM
INSTALLATION
HAZARDS

LOW TOXIC
EMISSIONS
IN USE

RESOURCE RECOVERY

OTHER

RESEARCH &
EDUCATION
PROGRAMS

PRODUCT
Patterned wool broadloom carpet

MANUFACTURER
H&I Carpet Corporation of Canada
231 Rowntree Dairy Rd.
Woodbridge, ON L4L 8B8
Tel (416) 850-1700   Fax (416) 850-1708

RELATED PRODUCTS
Adhesives, pp. 105–6, 116–23
Undercushion, pp. 55, 76–78

DESCRIPTION
DAMASK, MOIRE, TRELLIS, FLOWERS: Semiworsted wool, cut pile or loop construction. Low shedding, low mass Wilton carpet with jute and cotton backing. 2-tone patterning. Available in custom color. Width: 27".

Damask is compostable.

MANUFACTURER'S STATEMENT
More than 75% of H&I's business is writing specifications for custom contract carpet applications, using environmental and performance criteria.

H&I takes an active role in promoting environmental standards and raising awareness among design professionals. They have sponsored extensive research and education programs on the status of the carpet industry.

PRODUCT
Residential quality wool broadloom carpeting

MANUFACTURER
Nouwens-Bogaers BV (Holland)
Imported by Colin Campbell & Sons
202-1717 W. 5th Ave.
Vancouver, BC V6J 1P1
Tel (604) 734-2758

RELATED PRODUCTS
Adhesives, pp. 105–6, 116–23
Undercushion, pp. 55, 76–78

PRODUCTION

SUSTAINABLY ACQUIRED, OR RENEWABLE RESOURCE

IN-PLANT ENERGY EFFICIENCY & RECYCLING

LOW EMISSIONS PLANT

DESCRIPTION
BASSANO: 100% pure new wool face yarn, with jute backing and nontoxic latex adhesive. Variegated 2-tone color, Berber style construction. For use in heavy duty applications. Width: 13'2" (400cm). 4 colorways available.

PALLAS: 100% pure new wool face yarn, with jute backing and nontoxic latex adhesive. Variegated color, Berber style construction. For use in heavy duty applications. Width: 13'2" (400cm). 8 natural colorways available.

PACKING / SHIPPING

MANUFACTURER'S STATEMENT
Nouwens-Bogaers BV is a member of a group of European carpet manufacturers taking a stand on environmental issues.

Recently, processes have changed and they have invested in new, state-of-the-art machinery. Dye baths are recycled, reducing emissions by 50%, and remaining acids are neutralized. 40%–45% less wool is dyed each year. Although flocks are dyed in dark tones, the quantities of actual wool dyed are reduced. Dyed and undyed yarns are blended, producing a 2-tone effect. The yarn is washed in a closed-circuit process, to reduce emissions.

Nontoxic latex adhesive on backing has safe calcium carbonate additives.

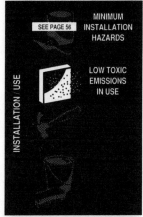

INSTALLATION / USE

SEE PAGE 56

MINIMUM INSTALLATION HAZARDS

LOW TOXIC EMISSIONS IN USE

RESOURCE RECOVERY

OTHER

PRODUCTION

SUSTAINABLY
ACQUIRED, OR
RENEWABLE
RESOURCE

LOW
EMISSIONS
PLANT

PACKING / SHIPPING

SEE PAGE 56

MINIMUM
INSTALLATION
HAZARDS

INSTALLATION / USE

RESOURCE RECOVERY

OTHER

PRODUCT
Residential quality wool broadloom carpeting

MANUFACTURER
Tintawn (Ireland)
Imported by

Tintawn Carpets
964 Third Ave
New York, NY 01022
Tel (212) 355-5030
Fax (212) 644-8359

Colin Campbell & Sons
202-1717 W. 5th Ave.
Vancouver, BC V6J 1P1
Tel (604) 734-2758
Fax (604) 734-1512

RELATED PRODUCTS
Adhesives, pp. 105–6, 116–23
Undercushion, pp. 55, 76–78

DESCRIPTION
IRISH COLLECTION: 100% pure new wool, Wilton woven construction. Cotton/jute warp and polypropylene weft backing with synthetic latex finishing.

Donegal: cut and loop pile construction, 15 colors available. *Corrib, Blarney Castle, Killarney, Waterford Weave, Shannon Shear:* cut and loop pattern construction. *Tralee:* 2-tone cut and loop pattern construction, 5 colorways available. 4 matching border patterns are available.

MANUFACTURER'S STATEMENT
Tintawn has been producing natural wool, sisal, and jute fiber floor covering products in Ireland for over 100 years. Continual upgrading of operations and maintenance of the original mill demonstrate their dedication to a traditional industry in the face of the popularity of synthetic products. While Tintawn operates a mill that produces natural and synthetic blends, their reputation rests on the quality of their wool and jute lines.

Recently, ownership of the company has been secured from the original offshore parent company by a consortium of employees in management positions.

MANUFACTURER

Van Besouw / KVT (Holland)
Imported by

Van Besouw BV/KVT International
Ltd.
240 Peachtree St., #4G5
Atlanta, GA 30303
Tel (404) 525-6643

Colin Campbell & Sons
202-1717 W. 5th Ave.
Vancouver, BC V6J 1P1
Tel (604) 734-2758
Fax (604) 734-1512

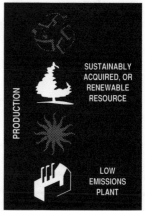

SUSTAINABLY ACQUIRED, OR RENEWABLE RESOURCE

PRODUCTION

LOW EMISSIONS PLANT

RELATED PRODUCTS

Adhesives, pp. 105–6, 116–23
Undercushion, pp. 55, 76–78

DESCRIPTION

SERIES 1405: 100% pure new wool, Wilton-woven style with jute backing. Yarns bonded to carpet face with high quality synthetic latex adhesive. Mothproofing treatment. Solid color, multilevel twist pile. Width: 13'2" (400 cm), suitable for heavy duty traffic. 6 stock colors, custom colors available.

SERIES 1406: 100% pure new wool. Backing bonded to carpet face with high quality synthetic latex adhesive. Mothproofing treatment. Width: 13'2" (400 cm), suitable for heavy duty traffic. 8 colors available.

SERIES 1407: 100% pure new wool, Wilton-woven style with jute backing. Yarns bonded to carpet face with high quality synthetic latex adhesive. Mothproofing treatment. Multicolor, multilevel twist pile. Width: 13'2" (400 cm), suitable for heavy duty traffic. 6 colorways available.

SERIES 1501 & 1502: 100% pure new wool, Axminster with jute backing. Yarns bonded to carpet face with high quality synthetic latex adhesive. Mothproofing treatment. Multicolor fleck-patterned cut pile. Width: 13'2" (400 cm), suitable for heavy duty traffic. 36 colorways.

SERIES 4401: 50% pure wool, 50% goat's hair broadloom carpet with jute backing yarns bonded to carpet face with high quality synthetic latex adhesive. Mothproofing treatment. High-low loop construction. Width: 13'2" (400 cm), suitable for heavy duty traffic, but not contract use. 16 colors available.

PACKING / SHIPPING

SEE PAGE 56

MINIMUM INSTALLATION HAZARDS

LOW TOXIC EMISSIONS IN USE

INSTALLATION / USE

MANUFACTURER'S STATEMENT

Van Besouw is a member of G.U.T., a European group of manufacturers committed to labeling textile floor covering products based on independent emissions testing by the German Research Institute for Carpet. Participants also pledge to keep their facilities on the leading edge of technology, in regards to air and water pollution controls.

Mothproofing treatment is performed to conform to Wool Bureau standards.

RESOURCE RECOVERY

OTHER

PRODUCTION

SUSTAINABLY ACQUIRED, OR RENEWABLE RESOURCE

LOW EMISSIONS PLANT

PACKING / SHIPPING

SEE PAGE 56

MINIMUM INSTALLATION HAZARDS

LOW TOXIC EMISSIONS IN USE

INSTALLATION / USE

RESOURCE RECOVERY

OTHER

PRODUCT
Residential quality cotton and linen broadloom carpeting

MANUFACTURER

Van Besouw BV/KVT International Ltd.
240 Peachtree St., #4G5
Atlanta, GA 30303
Tel (404) 525-6643

Colin Campbell & Sons
202-1717 W. 5th Ave.
Vancouver, BC V6J 1P1
Tel (604) 734-2758
Fax (604) 734-1512

RELATED PRODUCTS
Adhesives, pp. 105–6, 116–23
Undercushion, pp. 55, 76–78

DESCRIPTION

SERIES 3801: 100% cotton face yarn, Raschel-knitted with nonwoven synthetic fabric backing bonded to carpet face with high quality synthetic latex adhesive. Solid color, loop face pile construction. Width: 13'2" (400 cm), suitable for heavy duty traffic. 34 colors available.

SERIES 3802: 100% cotton face yarn, Raschel-knitted with nonwoven synthetic fabric backing bonded to carpet face with high quality synthetic latex adhesive. Solid color, loop construction, tip-sheared face pile. Width: 13'2" (400 cm), suitable for heavy duty traffic. 30 colors available.

SERIES 3803: 100% cotton face yarn, Raschel-knitted with nonwoven synthetic fabric backing bonded to carpet face with high quality synthetic latex adhesive. 2-color, high-low loop face pile construction. Width: 13'2" (400 cm), suitable for heavy duty traffic. 15 colorways available.

SERIES 3805: 100% cotton face yarn, Raschel-knitted with nonwoven synthetic fabric backing bonded to carpet face with high quality synthetic latex adhesive. 2-tone yarn, twisted loop face pile construction. Width: 13'2" (400 cm), suitable for heavy duty traffic. 15 colorways available.

SERIES 5401: 100% linen face yarn, with 100% linen backing bonded to carpet face with high quality synthetic latex adhesive. Width: 13'2" (400 cm), suitable for heavy duty traffic. 7 colors available.

MANUFACTURER'S STATEMENT

Van Besouw is a member of G.U.T., a European group of manufacturers committed to labeling textile floor covering products based on independent emissions testing by the German Research Institute for Carpet. Participants also pledge to keep their facilities on the leading edge of technology, in regards to air and water pollution controls.

PRODUCT
Residential and commercial natural fiber (sisal, wool, seagrass, and coir)

PRODUCTION — SUSTAINABLY ACQUIRED, OR RENEWABLE RESOURCE

PACKING / SHIPPING

INSTALLATION / USE — SEE PAGE 56 — MINIMUM INSTALLATION HAZARDS — LOW TOXIC EMISSIONS IN USE

RESOURCE RECOVERY

OTHER

MANUFACTURER
Merida Meridian Inc.
360 W. Genessee St.
Syracuse, NY 13201-1071
Tel (315) 422-4921

RELATED PRODUCTS
SISAL FIBER PROTECTION: Petroleum-based spray containing polymers (and trichlorethylene) that seal fibers against water spotting, soils, and stains. Helps keep sisal's original colors lightfast. Also for use with coirs and wools.

DESCRIPTION
MAYATEX, TASITWEED: 100% woven sisal floor covering lines, ranging from finer to rougher structures. Natural and dyed or patterned sisal, with a solid natural latex backing. Width: 12'5" or 13'1". Face yarn weights vary with style of weave. Mayatex has Class A flammability rating, Tasitweed has Class B flammability rating. Available in a variety of colors.

LLAMA LOOP: 100% tufted sisal floor covering with a solid natural latex backing. Width: 13'1"; face yarn weight: 45 oz. per square yard. Class B flammability rating. Available in variety of colors.

BERGE, SINGAPORE, AFRICAN WEAVE: 100% woven coir (57% coco, 43% sisal) floor covering with a solid natural latex backing. Width: 13'1" (also available in tiles); weight: 54 and 47 oz. per square yard, respectively. Both have Class A flammability rating. Available in variety of colors.

KARUNA: 100% woven coir floor covering with a solid natural latex backing. Width: 13'1"; 72 oz. per square yard. Available in 3 colors.

SEAGRASS: 100% seagrass flat-weave floor covering with a solid natural latex backing. Width: 13'1"; 67 oz. per square yard. Natural color only.

TOURNAI: 50% wool, 50% sisal floor covering with a solid natural latex backing. Width: 13'1"; 72 oz. per square yard. Available in 4 colors: charcoal, ecru, sand, and camel.

MANUFACTURER'S STATEMENT
Merida Meridian imports only natural fiber products which have minimal environmental impact. Yarns are spun and woven on Jacquard looms in Europe.

A range of custom-designed, patterned sisal area carpets is offered. Artists hand-paint original designs or hand-screen patterns from an in-house pattern collection.

95% of Merida Meridian's market is in loose-laid area carpets. For wall-to-wall installations, adhesive is recommended to deal with the variety of conditions encountered in sisal installation, such as high humidity, rough, and porous surfaces.

NOTE
Sealers and adhesives are toxic products, particularly during installation. Alternative low toxicity adhesives are listed on pp. 116–23. Low toxicity soil and water repellents may be available, although some experimentation and consultation with manufacturers may be required.

PRODUCTION

SUSTAINABLY ACQUIRED, OR RENEWABLE RESOURCE

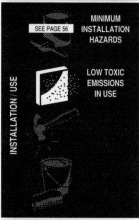

PACKING / SHIPPING

SEE PAGE 56

MINIMUM INSTALLATION HAZARDS

LOW TOXIC EMISSIONS IN USE

INSTALLATION / USE

RESOURCE RECOVERY

FAIR BUSINESS PRACTICES

OTHER

PRODUCT
Residential and commercial natural fiber (sisal, wool, seagrass, and coir)

MANUFACTURER
Ruckstuhl (Switzerland)
Imported by

Baumann Fabrics
302 King St. East
Toronto, ON M5A 1K6
Tel (416) 869-1221
Fax (416) 869-0428

Jack Lenor Larsen
41 E. 11th St.
New York, NY 10003-4685
Tel (416) 869-1221
Fax (416) 869-0428

DESCRIPTION
CARRÉ: 68% coir, 32% sisal in woven twill construction, with 8" × 8" woven pattern repeat. 79 oz. per square yard, unbacked. Roll width: 6'7"; roll length: 98'. 6 colorways available. Also available with PVC backing.

TRANSVERSE, CHEVRON: 43% coir, 57% sisal in woven twill and herringbone construction, respectively. 79 oz. per square yard, unbacked. Roll width: 6'7"; roll length: 98'. 5 colorways available. Also available with PVC backing.

RAJAH: 100% sisal in woven, bouclé rep weave. 58 oz. per square yard, unbacked. Roll width: 6'7"; roll length: 90'. 5 colors available. Also available with PVC backing.

MANILA: 100% sisal rep pattern. 79 oz. per square yard, unbacked. Roll width: 6'7"; roll length: 98'. 5 colors available. Also available with PVC backing.

CALICUT, TONGA: 100% coir in woven rep pattern. 99 oz. per square yard, unbacked. Roll width: 6'7"; roll length: 98'. Also available in 19¾" square tiles. Calicut available in 2 colors, Tonga available in 7 colors. Also available with PVC backing.

PASSAGE: 100% coir in woven rep pattern. 66 oz. per square yard, unbacked. Roll widths: 27½", 35½", and 47¼"; roll length: 164'. Available in 4 colors. Also available with PVC backing.

BRUSH UP: 100% coir in woven velour construction. 248 oz. per square yard, with PVC backing. Total height 1". Available in 7 colors.

VERANDA MATS: 100% woven coir in 4 patterns/colorways. Sizes 74" × 127", 78" × 118" and 110" × 165".

MANUFACTURER'S STATEMENT
A family-run business in its fourth generation, Ruckstuhl has intentionally retained a traditional outlook on the use of natural materials and prefers to pursue the quality of craftsmanship, as opposed to growing and expanding their facilities. The company does not make their materials to be indestructible, so that they will eventually break down when disposed of. Ruckstuhl views natural fibers as a precious raw material, so that the quality must not be compromised when converting them.

At stake is the livelihood of suppliers in an already hard-pressed world market. They believe that through insistence on quality, selective buying and the ensuing success, suppliers will strive to ensure a high quality and be able to continue to export the products of their homelands.

MANUFACTURER

Tintawn (Ireland)
Imported by

Tintawn Carpets	Colin Campbell & Sons
964 Third Ave	202-1717 W. 5th Ave.
New York, NY 01022	Vancouver, BC V6J 1P1
Tel (212) 355-5030	Tel (604) 734-2758
Fax (212) 644-8359	Fax (604) 734-1512

RELATED PRODUCTS

Adhesives, pp. 105–6, 116–23

DESCRIPTION

MANDALAY JUTE: 100% jute, flat woven construction with synthetic latex backing. 2-tone pattern, 10 colorways available.

BLENHEIM "HERRINGBONE" JUTE: 100% jute, flat woven construction with synthetic latex backing. 2-tone pattern, 10 colorways available.

MANUFACTURER'S STATEMENT

Tintawn has been producing natural wool, sisal, and jute fiber floor covering products in Ireland for over 100 years. Continual upgrading of operations and maintenance of the original mill demonstrate their dedication to a traditional industry in the face of the popularity of synthetic products. While Tintawn operates a mill that produces natural and synthetic blends, their reputation rests on the quality of their wool and jute lines.

Recently, ownership of the company has been secured from the original offshore parent company by a consortium of employees in management positions.

PRODUCTION

SUSTAINABLY ACQUIRED, OR RENEWABLE RESOURCE

PACKING / SHIPPING

SEE PAGE 56 MINIMUM INSTALLATION HAZARDS

LOW TOXIC EMISSIONS IN USE

INSTALLATION / USE

RESOURCE RECOVERY

OTHER

CARPETING
Nylon

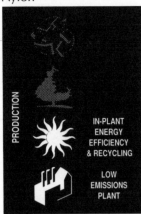

PRODUCTION

IN-PLANT
ENERGY
EFFICIENCY
& RECYCLING

LOW
EMISSIONS
PLANT

PACKING / SHIPPING

INSTALLATION / USE

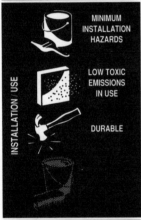

MINIMUM
INSTALLATION
HAZARDS

LOW TOXIC
EMISSIONS
IN USE

DURABLE

RESOURCE RECOVERY

REUSABLE,
SALVAGEABLE

OTHER

RESEARCH &
EDUCATION
PROGRAMS

PRODUCT
Commercial quality modular nylon carpet system

MANUFACTURER
Collins & Aikman (Floor Coverings Division)
311 Smith Industrial Blvd.
Dalton, GA 30722-1447 / Tel (404) 259-9711, (800) 241-4085

DESCRIPTION
POWERBOND RS: Continuous-filament DuPont Antron BCF dense yarn bonded to closed-cell PVC backing. Available in 6′ wide rolls and 18″ modular tiles. The backing has a built-in, factory-dried, microencapsulated tackifier. A protective plastic membrane is stripped and bonded to the floor without the use of wet adhesives. Allows easy, fast removal. Vinyl cushion extends carpet life, improves thermal and acoustical properties, and won't delaminate. Cushion is produced without CFCs or HCFCs.

Powerbond RS is guaranteed against watermarking, with the incorporation of Symtex, a fiber construction method. The closed-cell backing creates a moisture barrier that isolates liquids to the surface for easy cleanup. Yarn has superior soil hiding qualities, and provides easy maintenance. Chemical-welded seams, with urethane, provide durable, invisible, nonfraying edges and seams.

By eliminating the need for wet adhesives, offgassing of hazardous fumes is avoided. The closed-cell vinyl backing provides a barrier against radon seepage and asbestos release from worn floor tile. Airborne particles are reduced through the use of continuous-filament yarn and proprietary construction techniques. The heat-bonding technique, adhering the face pile to the backing, eliminates the need for latex.

MANUFACTURER'S STATEMENT
Collins & Aikman developed Powerbond over the last 20 years, in response to concerns over sick building syndrome, with the recent innovation of the releasable systems (dry adhesive). Powerbond is a low emission product and Collins & Aikman insist that suppliers eliminate carcinogens and toxins in their raw materials. Collins & Aikman offer a 15-year nonprorated warranty.

Recycling programs in-plant and in-office include requiring suppliers to provide reusable totes and drums; developing reusable cardboard containers and sending all tubes and scrap yarns to recycling centers; comprehensive office recycling programs, including paper, plastic, aluminum, cardboard, and yarn. Selvage wastes are recycled into plastic byproducts. Collins & Aikman are also in the process of implementing procedures to assure the proper disposal of their polyethylene membrane (from the material backing): specially designed bags will be supplied with the product so that this material can be recycled or disposed of properly.

Hot water is captured from many sources inside the plants to be used in the dye baths, reducing the energy requirements for this process. They have initiated a self-imposed environmental audit program called CLEAR (Confirm and Lower Environmentally Adverse Residuals) to remove impurities from their manufacturing facilities and their end product.

For every Powerbond RS order received, Collins & Aikman will plant a Colorado blue spruce in the Sierra Nevada National Forests.

PRODUCT
Residential quality woven nylon carpet series

MANUFACTURER
Crossley Carpet Mills Ltd.
435 Willow St., P.O. Box 745
Truro, NS B2N 5G2
Tel (902) 895-5491   Fax (902) 893-4779

RELATED PRODUCTS
ROBERTS EARTHBOND 7000, p. 120

DESCRIPTION
EMPEREAU CLASSIC: Woven cut and buried loop styling. Monsanto Acrilan Plus with integral stain resistant finish, 60 oz. per square yard. Carries the Crossley Lifetime Wear Warranty. Available in 25 colors.

REGATTA POINT: Woven cut and loop styling. Monsanto Acrilan Plus with integral stain resistant finish, 48 oz. per square yard. Carries the Crossley Lifetime Wear Warranty. Available in 18 colors.

SAND PATTERNS (a collection of four patterns—Collage, Swirl, Pebbles, Harmony): Woven loop pile construction, eliminating problems of shading and watermarking. Monsanto Acrilan Plus with certified resistance to stain, soil, static, matting, and fading. 48 oz. per square yard. Carries the Crossley Lifetime Wear Warranty. Available in 3 colors.

SEASHADOWS: Woven cut and loop pile construction, eliminating problems of shading and watermarking. Monsanto Acrilan Plus with certified resistance to stain, soil, static, matting, and fading. 60 oz. per square yard. Carries the Crossley Lifetime Wear Warranty. Available in 18 colors.

MANUFACTURER'S STATEMENT
Crossley's Crossweave weaving process contributes to longer appearance retention and thus an extended life cycle for the product.

Recently awarded provincial government recognition for energy efficiency for the installation of a radio frequency dryer. Recycled water is used for dyeing, and plans are under way for conversion to waste-to-fuel energy.

Crossley has initiated a product sample program which allows flexibility, and the ability to reuse architectural folders after a product is discontinued or recolored. Unlike before, where folders were discarded as a product was changed, or a folder was damaged, this new system will provide continuity in layout, and replaceable samples.

Crossley has in place several programs for employees: social support, profit sharing, pension plan, and fitness facilities.

PRODUCTION

IN-PLANT ENERGY EFFICIENCY & RECYCLING

PACKING / SHIPPING

MINIMUM, RECYCLED, RECYCLABLE PACKAGING

SEE PAGE 56

MINIMUM INSTALLATION HAZARDS

INSTALLATION / USE

DURABLE

RESOURCE RECOVERY

OTHER

FAIR BUSINESS PRACTICES

IN-PLANT
ENERGY
EFFICIENCY
& RECYCLING

PRODUCTION

MINIMUM,
RECYCLED,
RECYCLABLE
PACKAGING

PACKING / SHIPPING

SEE PAGE 56

MINIMUM
INSTALLATION
HAZARDS

DURABLE

INSTALLATION / USE

RESOURCE RECOVERY

FAIR
BUSINESS
PRACTICES

OTHER

PRODUCT
Commercial quality woven nylon carpet series

MANUFACTURER
Crossley Carpet Mills Limited
435 Willow St., P.O. Box 745
Truro, NS B2N 5G2
Tel (902) 895-5491   Fax (902) 893-4779

RELATED PRODUCTS
ROBERTS EARTHBOND 7000, p. 120

DESCRIPTION
AKARI: Woven cut and loop pile construction. 100% BASF Zeftron BCF nylon, 32 oz. per square yard. Polypropylene warp and weft, with latex finishing. Suitable for extra-heavy commercial traffic. Geometric texture and color blend patterning. Available in 13 colorways. Custom color available for a minimum order.

CALCULATIONS: Woven cut and loop pile construction. 100% BASF Zeftron BCF nylon, 32 oz. per square yard. Polypropylene warp and weft, with latex finishing. Suitable for extra-heavy commercial traffic. Combination of solid and heathered yarns. Available in 12 colorways. Custom color and weight available for a minimum order.

CHECKWEAVE: Woven cut and loop pile construction. 100% DuPont Antron Legacy nylon, 24 oz. per square yard. Polypropylene warp and weft, with latex finishing. Tone-on-tone checkerboard patterning. Available in 13 colorways. Custom color and weight available for a minimum order.

DYNAMIC DOTS: Woven cut and loop pile construction. 100% DuPont Antron Legacy nylon, 34 oz. per square yard. Polypropylene warp and weft, with latex finishing. Random colored elements in solid field. Available in 8 colorways. Custom color available for a minimum order.

BLUEPRINTS COLLECTION (Ecole, Watershed, Runway, Runway II, Milestone): Woven textured loop pile construction, 30–36 oz. per square yard. Polypropylene warp and weft, with latex finishing. Patterned and textured. Available in 7–9 colorways. Custom color available for a minimum order.

L'ATELIER COLLECTION (Hieroglyph, Symetrie, Couronne, Arabesque, Tirage, Tracerie): 6 patterns in cut and loop pile construction. 100% I.C.I. Timbrelle "S" nylon, 54 oz. per square yard. Polypropylene warp and weft, with latex finishing. Available in 14 colors. Custom color and weight available for a minimum order.

GLASSWORKS, JAVA: Woven cut and loop pile construction. 100% DuPont Antron Legacy nylon, 30 oz. per square yard. Polypropylene warp and weft, with latex finishing. Patterned. Available in 14 colorways. Custom weight available for a minimum order.

LISCOMBE: Woven cut and loop pile construction. 100% BASF Zeftron BCF nylon, 32 oz. per square yard. Polypropylene warp and weft, with latex finishing. Available in 8 stocked colors, with grid pattern. Custom color and weight for a minimum yardage.

MANUFACTURER'S STATEMENT
See p. 71.

PRODUCT
Residential and commercial quality polyester carpet from recycled fibers

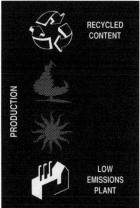

RECYCLED
CONTENT

PRODUCTION

LOW
EMISSIONS
PLANT

MANUFACTURER
Image Carpet
P.O. Box 5555
Armuchee, GA 30105
Tel (404) 235-8444, (800) 722-2504   Fax (404) 234-3464

RELATED PRODUCTS
Adhesives, pp. 105–6, 116–23
Undercushion, pp. 55, 76–78

DESCRIPTION
PRESERVATION: 100% Resistron (polyethylene terephthalate—PET) dense tufted, cut pile carpet with woven polypropylene backing. Width: 12', ¹⁄₁₀ gauge. 45 oz. per square yard, heavy commercial traffic rating. Available in 24 standard colors. Custom colors, weights, and widths available for large special orders. Product is treated with 3M Scotchguard (a topical finish) for additional soil resistance.

ENDLESS BEAUTY: Duratron (85% PET, 15% nylon blend) dense tufted cut pile carpet with woven polypropylene backing. Width: 12', ⅛ gauge. 75 oz. per square yard, residential quality. Available in 30 standard colors. Custom colors, weights, and widths available for large special orders. Product is treated with 3M Scotchguard Stain Release System (a permanent treatment) for additional stain resistance.

PACKING / SHIPPING

SEE PAGE 56

MINIMUM
INSTALLATION
HAZARDS

INSTALLATION / USE

MANUFACTURER'S STATEMENT
100% of yarn is extruded and spun from recycled polyethylene terephthalate (PET), derived from post-consumer soft drink and ketchup bottles (40 million pounds per year are diverted from landfills for this use).

Image eliminates the need for several thousand barrels of crude oil per year as raw materials for fiber. The fiber dyeing is performed in high pressure jet dye becks, eliminating the need for biphenyl ingredients as dye carriers. This method uses ⅔ the amount of water needed in conventional dyeing processes. The carpet finishing operations are conducted entirely with materials containing no formaldehyde.

Image will give assurance of satisfaction in the areas of wearability, texture retention, cleanability, and colorfastness.

RESOURCE RECOVERY

NOTE
Conventional stain repellents contain trichlorethylene, a chemical that emits toxic gases on installation. Use of stain repellents may not be appropriate for all applications.

OTHER

MINIMUM
INSTALLATION
HAZARDS

LOW TOXIC
EMISSIONS
IN USE

DURABLE

REUSABLE,
SALVAGEABLE

FAIR
BUSINESS
PRACTICES

PRODUCTION

PACKING / SHIPPING

INSTALLATION / USE

RESOURCE RECOVERY

OTHER

PRODUCT
Velcro-adhered backing system for synthetic broadloom carpeting

MANUFACTURER
Peerless International
Place Bonaventure, #1 Dawson, Mart D
Montreal, PQ H5A 1E8
Tel (514) 878-6800
Fax (514) 878-6829

RELATED PRODUCTS
Velcro strips
Adhesives, pp. 105–6, 116–23

DESCRIPTION
TAC FAST SYSTEM™: Solution-dyed yarns reaction-bonded to a polyurethane backing, for use as a backing system for broadloom carpets. Installed with dry method, using Velcro strips. Manufactured in 6' widths. Grid of Velcro strips are laid as required for differing traffic zones.

ENHANCER BACK™: Solution-dyed yarns reaction-bonded to a polyurethane backing, for use as a backing system for broadloom carpets. Direct glue-down installation method. Manufactured in 12' widths.

The use of solution-dyed yarn for all Peerless products minimizes excess dye stuff disposal and ensures that color is permanently locked into the construction of the fiber. The need for undercushion is eliminated with these backings. The polyurethane backing is dimensionally stable. The use of Velcro as a tacking system eliminates the use of wet adhesives, which emit volatile gases.

MANUFACTURER'S STATEMENT
The product can be modularized according to the needs of the particular job site, allowing broad goods where desired and subfloor access where required. Reuse potential is extremely high. Potential future reclaiming for conversion to stuffing and filling is also enhanced, as it is fairly inert.

Reaction-bonding the backing system to the face fiber fabric eliminates the need for a latex bonding agent. Independent toxicity testing concludes that there are minimal amounts of air contaminant emissions from the product itself.

PRODUCT
Carpet backing system for commercial carpeting in narrow roll widths and tiles

MANUFACTURER

Interface/Heuga
P.O. Box 1503
LaGrange, GA 30241
Tel (404) 882-1891, (800) 476-5556
Fax (404) 882-0500

Interface Flooring Systems
233 Lahr Dr., Box 1182
Belleville, ON K8N 5E8
Tel (613) 966-8090
Fax (613) 966-8817

RELATED PRODUCTS

HEUGATACK/GRIDSET: A pressure-sensitive, release agent adhesive for use in 6′ grid (4″ wide strip) only.

DESCRIPTION

SYSTEM SIX: Carpet backing system, available with all Interface and Heuga loop construction carpet products. Available on a custom order basis for any other product. Produced in 6′ widths only. Face fibers are thermally pigmented, and are several times more colorfast than yarn-dyed fibers.

Backing uses advanced chemical formulations of thermoplastic vinyl polymers for dimensional stability and resistance to delamination, raveling, and moisture penetration. Seams can be chemically welded for high seam integrity. Intersept®, a broad spectrum biocide inhibiting the growth of harmful bacteria and fungi, is an ingredient of the backing material.

Carpet is primarily a free-lay product, that is, it is inherently floor-hugging. Adhesive is recommended only in 4″ band under the seams and the peripheral edges, to create a holding grid. This reduces the quantity of adhesive required, and eliminates removal and reuse problems.

MANUFACTURER'S STATEMENT

System Six is fully constructed of synthetic materials. Minimal quantities of synthetic latex bonding agent are used, resulting in ⅓ to ⅕ less than EPA allowable VOC emissions. The synthetic construction is able to withstand harsh sanitation and cleaning procedures, resulting in minimal out-of-service periods.

Interface-Heuga Environmental Task Force has installed the following programs: paper and carton recycling; reduced waste in the manufacturing process by 50% in the past 2 years; increased the lengths of the carpet rolls in production, resulting in reduced seam waste; resale of waste yarn, saving 40,000 lbs. from landfill disposal. A return policy is being set up for carpet that has exhausted its useful life, to include it in the recycling program.

Interface-Heuga hosts *Lunch & Learn*, part of the Interface-Heuga Interlude Series of educational seminars and forums for designers, architects, specifiers and facility managers. They also publish and distribute an informational newsletter called *DesignNet*. Interface-Heuga is a partner of the Envirosense™ consortium who provide products manufactured with Intersept®.

NOTE

The addition of biocides to materials may not be an appropriate practice for many applications. Carpets should not be used in locations subject to chronic moisture and spills.

PRODUCTION

IN-PLANT ENERGY EFFICIENCY & RECYCLING

PACKING / SHIPPING

SEE PAGE 56 — MINIMUM INSTALLATION HAZARDS

INSTALLATION / USE

`LOW TOXIC EMISSIONS IN USE

DURABLE

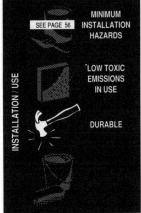

RESOURCE RECOVERY

REUSABLE, SALVAGEABLE

OTHER

RESEARCH & EDUCATION PROGRAMS

RECYCLED CONTENT

PRODUCTION

PACKING / SHIPPING

SEE PAGE 56

MINIMUM INSTALLATION HAZARDS

LOW TOXIC EMISSIONS IN USE

INSTALLATION / USE

RESOURCE RECOVERY

OTHER

PRODUCT
Synthetic fiber carpet undercushion

MANUFACTURER

Steiner-Liff Textile Products
(A Leggett and Platt Company)
P.O. Box 463
Nashville, TN 37202
Tel (800) 932-8517

Andy Miller & Associates
220 Midlake Blvd. S.E.
Calgary, AB T2X 1K6
Tel (403) 254-0309

DESCRIPTION

Nova Cushion is a synthetic fiber product with 80% recycled/converted fibers, side trim cuttings, and substandard materials obtained from carpet mills. Nova Cushion minimizes the need to restretch carpet, as the fibers "grip" the carpet backing to hold it in place.

NOVA SUPREME: Residential quality carpet undercushion, $7/16''$ thick. Available in 12' and 6' widths.

CONTRACT DELUXE: Commercial quality carpet undercushion, $3/8''$ thick. Available in 12' and 6' widths.

CONTRACT: Commercial quality carpet undercushion, $5/16''$ thick. Available in 12' and 6' widths.

SPECIAL CONTRACT: Commercial quality carpet undercushion, $5/16''$ thick, where a less dense product is desired. Available in 12' and 6' widths.

NOVA SLAB: Commercial quality carpet undercushion for double glue-down installation method. Available in 12' and 6' widths.

MANUFACTURER'S STATEMENT

Nova Cushion uses textile byproduct waste as a raw resource, efficiently utilizing material and energy that went into the original product.

PRODUCT
Commercial and residential quality recycled synthetic and natural fiber carpet undercushions

MANUFACTURER
Chris Craft Industrial Products Inc. (Waterford Division)
P.O. Box 70
Waterford, NY 12188
Tel (518) 237-5850

DESCRIPTION
COMMERCIAL LINE: 100% recycled jute from burlap bags used in the coffee bean and soybean industries. Bags are cleaned and sold through a broker, and acquired on the open market. Synthetic coating is applied to both sides, to contain the fibers. Available in 32, 40, 50, and 56 oz. per square yard, 6' and 12' widths. Suitable for higher traffic commercial use. R-value is approximately 2.5 per inch.

SUMMIT: 100% recycled synthetic fiber from carpet mill waste. With the material derived from a controlled source, the product contains a homogeneous mix of fibers, with consistency of color and quality. Residential qualities available in 22 and 28 oz. per square yard; commercial qualities in 32 and 38 oz. per square yard, 6' and 12' widths. R-value is approximately 2.5 per inch.

EXEL: 100% recycled fiber from textile industry waste. Waste materials are acquired from upholstery cutoffs and mill waste, and ground and reconstituted into the product. Residential qualities available in 18 and 24 oz. per square yard; commercial qualities available in 32 oz. per square yard, 6' and 12' widths. R-value is approximately 2.5 per inch.

MANUFACTURER'S STATEMENT
The products have been developed as an offshoot of the textile industry, to reduce solid manufacturing wastes, to increase cost and resource efficiency. The products are 99% recyclable and Chris Craft will take back their products at the end of their use for recycling. Products are not recommended for double-adhesive installation because of the inherent problems that render the products difficult to separate and recycle.

Chris Craft is active in an industry recycling initiative.

The synthetic products, Summit and Exel, contain a synthetic latex in their finishing, which amounts to 5% of the content of the overall product. The Commercial Line currently uses a synthetic coating, but Chris Craft is in the process of developing a natural rubber coating in order to produce a fully natural product. This improvement is expected to be in production very soon.

RECYCLED CONTENT

PRODUCTION

PACKING / SHIPPING

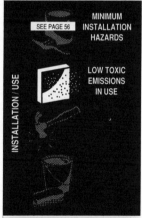

SEE PAGE 56

MINIMUM INSTALLATION HAZARDS

LOW TOXIC EMISSIONS IN USE

INSTALLATION / USE

RESOURCE RECOVERY

RECYCLABLE

OTHER

RESEARCH & EDUCATION PROGRAMS

RECYCLED
CONTENT

PRODUCTION

PACKING / SHIPPING

SEE PAGE 56

MINIMUM
INSTALLATION
HAZARDS

LOW TOXIC
EMISSIONS
IN USE

DURABLE

INSTALLATION / USE

RESOURCE RECOVERY

OTHER

PRODUCT
Commercial quality recycled rubber undercushion

MANUFACTURER

Dura Undercushions Ltd.
8525 Delmeade Rd.
Montreal, PQ H4T 1M1
Tel (514) 737-6561
Fax (514) 342-7940

Eastern U.S.:
Tel (908) 530-1155
Fax (908) 291-1277
Call (514) 737-6561 for
other U.S. distributors.

RELATED PRODUCTS

Adhesives, pp. 105–6, 116–23

DESCRIPTION

All products are made up of 56% rubber (90% of which is from recycled material), 6% oil extenders and plasticizer, 18% carbon black, 8% curing compounds, and 12% backing material.

DURACUSHION: Manufactured with a cellular structure made of rubber granules from ground tire scrap, bonded with latex, with jute backing. Installation by conventional stretching methods. 54″ wide roll.

DURACUSHION WIDE (for area rugs): Manufactured with a cellular structure made of rubber granules from ground tire scrap, bonded with latex, with jute backing. Piece material 9′ × 12′ dimensions.

DURALUX: Open cellular rubber, reinforced with solid rubber particles bonded to fiberglass/cellulose backing. Recommended for heavy commercial use. Installation by double glue-down system. 54″ wide.

SUPER DURA: Cellular structured undercushion made from ground tire scrap rubber granules, bonded with latex (90% of the rubber content of the product is from recycled material), with a polyester or jute backing. 54″ wide roll.

MONOSLAB I, MONOSLAB II: Cellular structured undercushions made of rubber granules from ground tire scrap, bonded with latex (90% of the rubber content of the product is from recycled material). Monoslab I has a jute top layer, while Monoslab II has top and bottom jute layers. 54″ wide roll. Reinforced and recommended for high traffic use.

MANUFACTURER'S STATEMENT

Dura Undercushion Ltd.'s corporate policy is to improve on their standard of 90% recycled rubber material, and thereby help the world strive for a goal of zero pollution.

FLOORING

• WOOD • STONE • CERAMIC • RESILIENT •

WOOD FLOORING

Solid wood flooring is a valued flooring option, and is considered a renewable resource when purchased from suppliers who practice responsible logging. A wood floor does not collect dust and soil as do carpets, and is easy to clean, making it an excellent choice for people with allergies to dust and dust mites.

SOLID PLANKING AND STRIPS

SOFTWOODS

Softwoods have not been used much for flooring recently. However, many older homes still retain their original fir or pine flooring. Though the high quality of wood required for softwood floors is difficult to find, demolition companies, in the past few years, have helped by selling salvaged fir and pine flooring. Renovators frequently remove carpet and resilient flooring, and restore the original wood floor. Old weathered barn wood is also occasionally available for flooring, from salvage yards. It provides a country look, but is more difficult to maintain. It takes some luck, research, and persistence to locate these original materials. Both the concept of reuse and the effect, though, are well worth the effort.

HARDWOOD

Hardwood flooring comes largely from domestic producers, with very little tropical wood used. The most common North American species used are red and white oak, maple, and sometimes birch. Imported beech (from Scandinavia) and, more recently, eucalyptus, also known as gum (from Australia) are also available. North American beech is one of the most depleted hardwood species, and should be avoided.

Solid dimensional stock is required for plank-type flooring installation, and the material can range from ¾" × 3" to 1½" × 6". This use requires high grade timber, from slow growing trees, and often the demand cannot be met by using only plantation harvested trees. Suppliers with responsible logging practices are few, and you should ask about the source of the material you purchase.

Another form of solid hardwood flooring is a thin strip (usually $3/8'' \times 2''$), which is nailed to a subfloor through the face, and the nail heads filled. It uses less wood, but can be sanded fewer times. The great advantage of a solid wood floor is its ability to be refinished (optimally every five to ten years). This ensures a long life and, thus, an appropriate use of the resource. (See the FINISHES section for sealing and finish options.)

COMPOSITE WOOD FLOORING

Many flooring systems are available that are composed of a thick veneer of finished hardwood laminated to a plywood or particle board backing, and packaged in tiles. These products are available in many plank and parquet patterns, colors and decorative borders. The overall product is usually thin enough to be able to be installed on a concrete subfloor. These flooring systems are generally prefinished, so they do not require sanding or finishing on site, unlike solid planking. They are, however, glued to the subfloor, exposing workers to emissions from toxic adhesives.

These flooring systems have two environmental benefits: value is confered on low grade wood in the plywood portion; and a high grade wood resource is extended efficiently by the use of veneer. Some manufacturers utilize reconstituted veneers, which are made from lower grade woods that have been machined and dyed, further lessening the environmental impact. This type of flooring is not able to be refinished, however, and the veneer can wear through in high traffic areas. It can be installed with low toxicity glues, reducing the exposure to tradespeople.

Refer to FURNITURE, pp. 203–8, for a more in-depth discussion of the environmental and health impacts of solid wood, plywoods, and veneers.

STONE AND CERAMIC FLOORING

Stone and ceramic flooring both offer a durable and easily maintained surface. They impart a solid, natural feeling to an interior, and, though nonrenewable, large quantities of the raw resources are readily available. Quarrying the material has a great impact on the land, altering drainage patterns, producing erosion, and transforming the landscape dramatically. Environmental restoration of a quarry requires filling, drainage, and planting as compensation for the environmental damage. This is rarely done fully, but partial restoration is now required in some areas.

STONE FLOORING

While general resource deposits of stone are vast, certain types of stone, quarried over centuries or in intense demand, have become in some cases almost completely depleted. A stone's reputation can create a worldwide demand, commanding extremely high prices for the little that is left. Since the desired color and texture are so closely linked to its location, a few quarries ship a lot of material very long distances. Though it has a durable finish once installed, stone (such as sandstone, slate, marble, and granite) is very fragile in transit. The risk of damage to a shipment and virtual waste of the entire load is great. In addition, the weight is so high that even local shipping is costly. All these elements are factored into the price, making stone a very expensive material, especially when imported.

Stone for interior finishing is available in irregular slabs (flagstone), uniformly cut and gauged slabs (with a thickness of ¾"), and uniformly cut and gauged tiles (with a thickness of approximately ⅜" and usually 12" × 12"). There is a wide selection of surface textures, such as polished, flamed, and pummeled, and all require mechanized processing, which adds to the energy consumption. North American quarries have high operating costs, but labor costs are low in Asia, southern Europe, and South America. This has resulted in exploitation of both the resource and the work force in these areas of the world.

The use of local materials is often praised by architects and individuals in that it helps to situate and integrate a building into its surroundings. The use of a resource from our backyard also fosters an attitude of conservation, and encourages rehabilitation of quarry sites. One concern with local stone is the possible contamination by radon, a naturally occurring radiation, which increases the risk of lung cancer. Check with local authorities, or your regional Environmental Protection Agency (EPA) or Environment Canada office to determine if radon is a concern in your area.

Similarly to ceramics, the natural surface of all but the most durable stone, such as granite, needs to be sealed to withstand soiling and staining. This sealer will have to be periodically renewed to maintain the surface. (See the FINISHES section for finishing options.)

CERAMIC FLOORING

Ceramics are fired clays, of which there are vast supplies. Their manufacture uses a great deal of energy (usually natural gas) in the firing process. Fluoride emissions from the kilns are the major cause of environmental damage inherent to the industry.

There are many grades, densities, and finishes of ceramics. Tiles are often glazed with fused minerals that have similar properties to the fired clay bodies, rendering the surface smooth and easy to clean. Unglazed tiles need a coat of sealer to achieve the same ease of maintenance, and will need periodic resealing. Floor tiles tend to be thicker than wall tiles because of their harder use.

Due to the permanent nature of tile installation, reuse of tiles is not practical. True recycling is impossible because, after the firing process, the clay material is changed at the molecular level, and cannot be broken down and made into tiles again. Some manufacturers are now utilizing waste glass and feldspar (a waste byproduct of the ceramics industry itself) as filler for floor tile bodies. Alternatively, used ceramics can be crushed and used as inert filler, or aggregate material as a base for sidewalk and road construction.

INSTALLATION METHODS

Installation methods are similar for both ceramic tiles and stone, as are the materials and compounds used, though stone is always set with mortar, and is usually placed on a concrete subfloor. Some emissions may occur from the use of sealers, waxes, or other finishes, but low toxicity sealers and safe, cement-based mortars are readily available. (See INSTALLATION MATERIALS for an outline of product types, and their advantages and disadvantages.)

Both stone and ceramic tile are highly chemically stable materials and will not deteriorate significantly with age. Although the processing of both has high energy consumption and produces some pollution, they are good choices overall from an environmental standpoint due to their durability and the wide availability of raw materials.

RESILIENT FLOORING

The following types of resilient flooring materials are available in both tiles and sheet goods. For added ease of maintenance, a clear sealer is generally recommended if it has not already been applied in the factory. There is a hierarchy of installation methods, from safest to least safe, with the choice of adhesive being the most important consideration. There are loose-laid, dry adhesive, low toxicity adhesive, and conventional latex- and solvent-based adhesive methods. (See INSTALLATION MATERIALS, pp. 105-7, for further information.) For most of the following materials, recycling or reuse possibilities are limited, or experimental. All are durable, with specific properties that can fulfill a wide range of needs.

CORK

Cork sheeting is compressed cork dust, generally obtained as a waste material from the bottle stopper industry. It is used as flooring, tackboards, and sometimes wall covering. The bark is peeled off cork oak trees on plantations in Portugal and Spain every eight to ten years, and yields a sustainable natural material that undergoes very little processing. Cork is a durable material with good thermal and acoustic properties, and it has a shock-resistant, elastic texture. It dissipates static electricity and is nonconductive.

LINOLEUM

Linoleum is a traditional material processed with natural, renewable ingredients (linseed oil, cork, wood dust, and dyes), that have been heat-cured. It has been made by the same methods for several generations. Linoleum is used for floors and counter or desk tops. The high linseed oil content imparts natural antibacterial properties, and renders the surface resistant to grease, oil, and diluted acids, as well as making it almost nonflammable. It is resilient, flexible, nonconductive, nonstatic, and extremely durable. Linoleums are now becoming available once more in a broad range of colors and patterns.

RUBBER

There are many products on the market, some from new, synthetic manufacture, and many from shredded, post-consumer sources. New rubber depletes petroleum reserves and causes serious air and water pollution. Rubber from recycled tires has two environmental benefits: that of utilizing an already manufactured material, and that of reducing the vast waste of discarded tires.

The final product is very resilient, shock-absorbing and sound-deadening—ideal for sports arenas, gymnasiums, recreation and fitness rooms, and play-grounds. However, rubber is highly flammable, and also produces continual minor offgassing. The odors are merely objectionable, though, and not hazardous.

VINYL

Vinyl is manufactured from petroleum and is available in two types: rigid tiles and flexible sheet goods. Of the two, the rigid material is more chemically and dimensionally stable, and thus more durable and less of a health risk. The soft plastic type contains plasticizers and sometimes foam, adding serious environmental and health impacts. In both cases, the manufacturing process causes serious air and water pollution at every stage of petroleum extraction, refining, and processing. The risk of chemical injury to workers is high at the time of manufacture, and the exposure of installers to the emissions from conventional adhesives can cause chronic illnesses.

INSTALLATION METHODS

Vinyl, cork, rubber, and linoleum are usually applied with a flooring adhesive. Though conventional adhesives are toxic, low toxicity alternatives are available for all applications. (See INSTALLION MATERIALS, p. 106.)

One advantage of the heavier natural linoleum sheets is that they can be laid dry, without adhesives, in small rooms like bathrooms or kitchens, and will not move. Heavy rubber goods can also be laid dry in recreation and fitness areas.

For larger floor areas, resilient sheet flooring is also joined at the seams by two permanent methods. The first, "solvent welding," is used only for synthetic materials. It requires the use of a toxic solvent to soften the material at the edge so that it can be joined to the next sheet. There are no safer alternatives to this solvent. The second is "heat welding," and it is used both for natural linoleums and synthetics. The edge is heated with an electric tool and a thin strip of the material is used to fuse the edges. This method produces hazardous fumes with synthetics, and only moderately irritating fumes with natural linoleums.

RECYCLING AND RECOVERY

Only dry-laid goods can be successfully salvaged for reuse. There is no functioning system in North America for recycling plastic flooring materials after use, but an experimental system has begun in Europe for recycling vinyl products.

SUSTAINABLY
ACQUIRED, OR
RENEWABLE
RESOURCE

MINIMUM
INSTALLATION
HAZARDS

LOW TOXIC
EMISSIONS
IN USE

DURABLE

SIMPLE,
NONTOXIC
MAINTENANCE

REUSABLE,
SALVAGEABLE

PRODUCT
Commercial, institutional, and residential strip wood flooring

MANUFACTURER

Junckers (Denmark)
Imported by

Junckers
Suite P, Bldg. 1
187 W. Oregethorpe Ave.
Placentia, CA 92670
Tel (714) 579-3188

Centaur Products Inc.
6855 Antrim Ave.
Burnaby, BC V5J 4M5
Tel (604) 430-3088
Fax (604) 430-1393

RELATED PRODUCTS

BLITSA STRONG 825: Low odor, water-based 2-component lacquer, used to maintain old presealed floors and for sealing sanded floors. Silk-matt sheen with high durability.

BLITSA, BLITSA MATT, FLOOR LACQUER 222: Low odor, water-based lacquer, touch dry in 1 hour, and walking dry in 2–3 hours. For interior use only. Floor Lacquer 222 will also seal cork floors, and untreated and previously varnished wood floors.

ISOLACQUER: 2-component isocyanate polyurethane lacquer, comprising matt lacquer and hardener. Silk-matt sheen with high durability and resistance to chemicals.

DESCRIPTION

JUNCKERS SPORTS FLOOR WITH CLIP FLOOR SYSTEM: Flooring available in solid beech, oak, and ash, ⅞″ (22mm) thick. Comes prefinished with a 2-part clear urethane (isocyanate) lacquer finish. The beech flooring undergoes a press-drying technique, which ensures a uniform moisture content in the wood.

A free-floating system designed to be installed over an existing floor or subfloor. Spring steel retainer clips are fitted to the undersides of the flooring boards, and a ½″ (12mm) rubber underlayment is used. The entire system elevates the floor level only 1½″ (34mm). Suitable for gymnasiums, sports halls, ballrooms, roller skating rinks, badminton courts, fitness centers, portable floors, squash and racquetball courts.

JUNCKERS RESIDENTIAL WOOD FLOORING WITH CLIP FLOOR SYSTEM: Solid beech, oak, and ash flooring, ⅞″ (22mm) thick. The oak and ash are kiln-dried, and the beech is press-dried. Prefinished with 2-part urethane that requires no waxing, and plastic vapor barrier laminated to underside of each strip. Various grades of woods are available, with different grain effects: prime, standard, flamey, and extra flamey.

MANUFACTURER'S STATEMENT

Junckers solid hardwood floors are prefinished in the factory, eliminating the need for wet-finishing in place, and frequent reapplication. Since several years usually elapse between lacquerings, a self-adhesive reminder note (indicating which product was used) is provided on the top of all Junckers lacquered products that can be stuck to the skirting of the floor. The clip floor system requires no adhesives, and is easily installed and removed for reuse.

Danish logging practices have had to become sustainable, with companies owning and logging their own forest.

PRODUCTION

PACKING / SHIPPING

INSTALLATION / USE

RESOURCE RECOVERY

OTHER

MANUFACTURER
Rodman Industries
P.O. Box 76
Marinette, WI 54143
Tel (715) 735-9541

RELATED PRODUCTS
Paints and stains, pp. 136–42, 144–65
Varnishes and sealers, pp. 142, 166–73
Oil finishes, pp. 184–86

DESCRIPTION
RESINFLOOR: High density flooring panel with water-resistant, phenol-based resins and blended hardwood fibers. Available in 4' × 8' × ¾" thick. Comes predrilled and countersunk for wood screws, 1" from perimeter and 8" on center, with two rows, 16" apart, 8" on center in interior of board.

Specially sanded for skid-resistant surface, for factory and warehouse floors. Material cannot delaminate or splinter, and has high impact and moisture resistance.

MANUFACTURER'S STATEMENT
Rodman Industries can consume 220 tons of sawdust per day, a byproduct of lumber and planing mills. Phenolic resins have no urea-formaldehyde (UF) and, since they do not emit significant quantities of formaldehyde, they are exempt from provisions of the HUD safety standard. Phenolics have a greater water and heat resistance than UF resins.

Resinfloor is almost twice as hard as oak flooring. It can reduce material cost by as much as 30 percent by eliminating the waste that occurs with hardwood strip flooring because of warpage, splits, and grain-matching losses.

PRODUCTION — SUSTAINABLY ACQUIRED, OR RENEWABLE RESOURCE

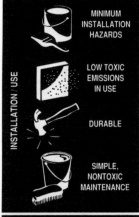

PACKING / SHIPPING

INSTALLATION / USE — MINIMUM INSTALLATION HAZARDS — LOW TOXIC EMISSIONS IN USE — DURABLE — SIMPLE, NONTOXIC MAINTENANCE

RESOURCE RECOVERY

OTHER

PRODUCTION

PACKING / SHIPPING

MINIMUM,
RECYCLED,
RECYCLABLE
PACKAGING

INSTALLATION / USE

SEE PAGE 81

MINIMUM
INSTALLATION
HAZARDS

LOW TOXIC
EMISSIONS
IN USE

DURABLE

SIMPLE,
NONTOXIC
MAINTENANCE

RESOURCE RECOVERY

OTHER

PRODUCT
Natural quarried limestone

MANUFACTURER

Gillis Quarries Ltd.
94 Wenzel St.
Winnipeg, MN R2C 2Z2
Tel (204) 222-8319
Fax (204) 222-7849

Weber Consulting
504 Montrose Rd.
St. Cloud, MN 56301
Tel (612) 251-1540, 255-5132
Fax (612) 259-9264

RELATED PRODUCTS
Mortar, pp. 104, 112

DESCRIPTION

TYNDALL STONE: A natural, quarried, medium density, light colored, mottled dolomitic limestone.

Cut Stone: custom cut and shaped to specific dimensions for a particular project. Uses: exterior and interior walls, columns, sills, steps, platforms, copings, flooring, table tops, roof tiles.

Random Ashlar: standard stock, precut coursed stone strips in various standard sizes and thicknesses. Uses: random facings for buildings, homes, fireplaces, chimneys, planters, garden and retaining walls, and as flagging for walks and patios.

Colors: *Buff,* a light creamy-beige with pastel brown mottling; *Gray,* a pale bluish-gray with gray-brown mottling.

Finishes: *Rubbed,* machine ground; *Sawn,* shows circular diamond tooth saw marks; *Bush hammered,* fine-grain textured surface; *Pointed,* rough textured, picked, or pebbled finish; *Split finish,* irregular rock-like split finish; *Rustic,* rugged texture by wedging stone apart. Sandblasting and grooves or patterns by sawing and grinding are possible.

Sizes: 90mm thickness standard for cut stone panels, 57mm available. Maximum 600mm without styolitic seams.

MANUFACTURER'S STATEMENT

Silica content is very low and negligible in most layers, ensuring minimal health hazards from cutting. Cutting waste is suitable for rubble fill. Fine cuttings have been approved as agricultural limestone.

Material is palletized on skids or returnable pallets with steel strap and, in some cases, shrinkwrap. Tyndall stone is not shipped internationally because, unlike granite, it cannot be cut thinly, with resultant high weight and freight costs.

In quarry pits which are exhausted, ledges are stabilized. Park development has been attempted, including stocking with trout. These have been successful physically, but not economically, partly because of the short summer season.

PRODUCT
Architectural ceramic tile and accessories

MANUFACTURER
Gail, Ag. (Germany)
Gail Representatives:
Western Canada (604) 879-7711   Eastern U.S. (908) 747-9797
Ontario (416) 293-3664   Western U.S. (714) 579-3080
Quebec (514) 384-5590   Southern U.S. (214) 243-6100
Maritimes (902) 420-0277

RELATED PRODUCTS
Adhesives, pp. 105–6, 116–23
Mortar, pp. 104, 112
Grout, pp. 104, 113-14
Sealers and oil finishes, 142, 173–74, 184–86

DESCRIPTION
KERA SPECTRUM: Extruded ceramic tiles, glazed and unglazed, in a full range of sizes and accessories.

KERA DESIGN AND KERA PLUS: Glazed dry pressed tiles, solid and patterned, in a full range of sizes and accessories.

KERA FORTE: Unglazed, dry pressed, porcelain-style floor tiles, available with textured, smooth, and polished surfaces. Accessories available.

KERA FLAIR: Super-thin and extra-large glazed wall tiles, solid and patterned, with 4 styles of borders available.

KERA SYSTEM: Full range of special-form fittings and accessories for swimming pools.

KERA CARE: Special ceramics for safety and direction-finding functions. Antislip glazed and antislip textured surfaces, in a limited range of sizes and colors.

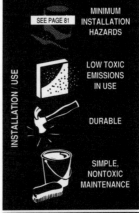

MANUFACTURER'S STATEMENT
Gail uses nontoxic glazes and color pigments.

Worked-out clay pits are filled and replanted as parks. In 1982, Gail was awarded a Gold Plaque and Special Distinction in a nationwide competition, entitled "Industry in the Landscape," for their recultivation of a clay pit which is protected today as a nature reserve.

Gail's in-plant waste reduction and comprehensive recycling programs exceed official German legislation requirements. Kilns use natural gas. Excess heat is used in drying kilns. All scrap material is recycled into the production process.

RECYCLED
CONTENT

PRODUCTION

PACKING / SHIPPING

SEE PAGE 81 | MINIMUM
INSTALLATION
HAZARDS

LOW TOXIC
EMISSIONS
IN USE

DURABLE

SIMPLE,
NONTOXIC
MAINTENANCE

INSTALLATION / USE

RESOURCE RECOVERY

OTHER

PRODUCT
Glass-based ceramic tile

MANUFACTURER
GTE Engineered Ceramics
(division of GTE Products Corporation)
1 Jackson St.
Wellsboro, PA 16901
Tel (717) 724-8322

RELATED PRODUCTS
    Adhesives, pp. 105–6, 116–23
    Mortar, pp. 104, 112
    Grout, pp. 104, 113-14

DESCRIPTION
PROMINENCE: Unglazed, glass-based ceramic floor tile. Fusion of glass and clay means that the tile is impervious and stain resistant with no surface porosity and does not require waxes, sealers, or coatings. Prominence is dimensionally consistent, slip and scratch resistant, and rated for high-traffic applications.

Available in 30 different colors, a range of monochromes and polychromes. Pigments mixed into the white body create a complete color infusion for full color saturation. Available in 8″ × 8″, 12″ × 12″, and trim pieces.

MANUFACTURER'S STATEMENT
GTE uses excess glass (mill waste) from their production of Sylvania light bulbs for the main ingredient of the tile body. Prominence is impervious and is unaffected by acids or alkalis.

MANUFACTURER
Stoneware Tile company
1650 Progressive Dr.
Richmond, IN 47374
Tel (317) 935-4760   Fax (317) 935-3971

RELATED PRODUCTS
Adhesives, pp. 105–6, 116–23
Mortar, pp. 104, 112
Grout, pp. 104, 113-14

DESCRIPTION
TRAFFIC TILE: Glass-bonded ceramic tile with color-saturated tile body. 19 standard, vibrant pure colors; unlimited range of custom colors available. The recycled glass used accepts color stain and reproduces the true color without distortion. This permits a level of clarity and brilliance that is impossible to achieve using the standard clay body, which has a tendency to diffuse color and shift it toward shades of brown.

Fire-sealed, fully modular, embossed back, wall, and floor tile for interior and exterior use. ⅜" thickness available in nominal sizes: 4" × 4", 4" × 8", 6" × 6", and 8" × 8", with trim pieces (round and square top, outcorners, bullnose and bullnose corners). Custom sizes, colors, and abrasive finish available on request.

CRAFTSMAN LINE: Commercial and residential glazed tile with low water absorption for interior and exterior use. Ceramic scrap from Traffic Tile and recycled glass together make up over 90% of raw material used in Craftsman tile body.

8 natural, mottled colors. Custom colors and sizes available. Softly formed edges, irregular surfaces for a handcrafted look, with antiskid finish available. Frost resistant. Nominal ½" thickness in 3 standard sizes: 4" × 4", 4" × 8" and 8" × 8". Countertop trims (sink trims include inside and outside corner trim, bullnoses, and accent pieces) available.

MANUFACTURER'S STATEMENT
Traffic Tile is composed of 70% recycled glass.

Both Traffic Tile and the Craftsman Line are clear-fired with "Dura Floor 11" resulting in a dirt-resistant, maintenance-efficient barrier.

Stoneware Tile has been manufacturing the glass-bodied tiles since 1988, using proprietary formulations first introduced in the 1970s.

RECYCLED
CONTENT

PRODUCTION

PACKING / SHIPPING

SEE PAGE 81

MINIMUM
INSTALLATION
HAZARDS

LOW TOXIC
EMISSIONS
IN USE

DURABLE

SIMPLE,
NON TOXIC
MAINTENANCE

INSTALLATION / USE

RESOURCE RECOVERY

OTHER

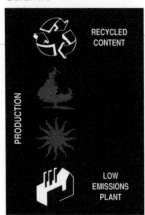

RECYCLED CONTENT

PRODUCTION

LOW EMISSIONS PLANT

PACKING / SHIPPING

SEE PAGE 81 — MINIMUM INSTALLATION HAZARDS

LOW TOXIC EMISSIONS IN USE

DURABLE

SIMPLE, NONTOXIC MAINTENANCE

INSTALLATION / USE

RESOURCE RECOVERY

OTHER

PRODUCT
Glazed porcelain pavers

MANUFACTURER
Summitville Tiles Inc.
P.O. Box 73
Summitville, OH 43962
Tel (216) 223-1511   Fax (216) 223-1414

RELATED PRODUCTS
Adhesives, pp. 105–6, 116–23
Mortar, pp. 104, 112
Grout, pp. 104, 113-14

DESCRIPTION
OCTAGON/DOT: Porcelain paver. 8″ octagons with 2″ dots are available in smooth and abrasive surfaces. *Imperva, Imperva Granite, Morganmates* and *Summitshades* styles.

IMPERVA: Heavy duty commercial glaze with high density, frostproof body. Available in smooth and abrasive surfaces, in 11 colors.

IMPERVA GRANITE: Impervious porcelain paver with granite-like glazed surface, frostproof body suitable for heavy duty commercial and residential applications. Available in smooth and abrasive surfaces, in 11 colors.

MORGANMATES: Impervious glazed porcelain pavers available with a smooth or abrasive surface, for indoor and outdoor use. Frostproof body with durable glazed surface, available in 11 colors.

SUMMITSHADES: Impervious glazed procelain pavers, with frostproof body and smooth or abrasive surface, for commercial and residential applications. Available in 6 shades/tones.

MANUFACTURER'S STATEMENT
Since 1989 Summitville has been producing porcelain pavers in a new state-of-the-art facility. The major body constituent uses a waste product of feldspar refining. Feldspar is used as a filler and a flux in tile bodies and glazes, reducing waste generated in the tile industry that has been traditionally landfilled. Summitville invested 3 years of research and development into the use of this waste as the main raw resource for tile production.

The result is a closed system for solid waste accumulation and reuse, consuming virtually all unprocessed wastes, and incorporating them into the body batching system.

MANUFACTURER
Corticeira Amorim
Imported by FloorEvery Natural Cork Flooring Products
865 Wall St.
Winnipeg, MN R3G 2T9
Tel (204) 772-4398   Fax (204) 775-1819

RELATED PRODUCTS
FLOOREVERY SEALER: Low toxic, waterborne barrier sealer, to maintain the lightness of the cork.

FLOOREVERY URETHANE: Low toxic latex coating.

FLOOREVERY ADHESIVE: Low toxic, waterborne contact adhesive.

DESCRIPTION
FLOOREVERY: Cork tiles consisting of 100% cork particles compressed with natural resins.

Standard sizes: $12'' \times 12'' \times \frac{3}{16}''$ or $\frac{5}{16}''$. Rolls up to 36" wide, and custom sizes available on special orders, with minimum quantities required. Available in 9 patterns which can be cut into several tile sizes or planked. The cork flooring may also be stained, using Pratt & Lambert tinting system.

MANUFACTURER'S STATEMENT
FloorEvery utilizes natural resins in the manufacturing process, resulting in a hypoallergenic product. FloorEvery has applied for certification under Environment Canada's Environmental Choice Program, for the Ecologo.

PRODUCTION

SUSTAINABLY ACQUIRED, OR RENEWABLE RESOURCE

PACKING / SHIPPING

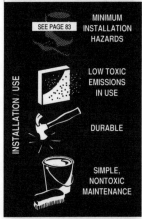

INSTALLATION / USE

SEE PAGE 83

MINIMUM INSTALLATION HAZARDS

LOW TOXIC EMISSIONS IN USE

DURABLE

SIMPLE, NONTOXIC MAINTENANCE

RESOURCE RECOVERY

OTHER

PRODUCTION

SUSTAINABLY
ACQUIRED, OR
RENEWABLE
RESOURCE

PACKING / SHIPPING

SEE PAGE 83   MINIMUM
INSTALLATION
HAZARDS

INSTALLATION / USE

LOW TOXIC
EMISSIONS
IN USE

DURABLE

SIMPLE,
NONTOXIC
MAINTENANCE

RESOURCE RECOVERY

OTHER

PRODUCT
Cork tile

MANUFACTURER

Dodge-Regupol Inc.
P.O. Box 989,
Laurel & Manor Streets
Lancaster, PA 17603
Tel (717) 295-3400, (800) 322-1923
Fax (717) 295-3414

Phoenix Floor and Wall Products
111 Westmore Dr.
Rexdale, ON M9V 3Y6
Tel (416) 745-4200
Fax (416) 745-4211

RELATED PRODUCTS

Varnishes and sealers, 166–74
Oil finishes, pp. 184–86
Waxes, pp. 187–190

DESCRIPTION

DODGE CORK TILE: Resilient flooring consisting of 100% compounded cork particles.

Standard sizes: 12″ × 12″ × ³⁄₁₆″ or ⁵⁄₁₆″. Rolls up to 36″ wide, and custom sizes available on special orders, with minimum quantities required.

Available in 3 shades: light random, medium random, and dark random. Available in 3 finishes: naturally unfinished, polyurethane coated, and standard wax coated.

MANUFACTURER'S STATEMENT

Dodge-Regupol utilizes the waste byproduct of the bottle cork industry in the Iberian Peninsula.

MANUFACTURER
Ipocork (Portugal)
Imported by
Ipocork USA
1280 Roberts Blvd., Suite 403
Kinnesaw, GA 30144
Tel (404) 421-9567

Eastern Canada:
Tel (514) 322-9110
Fax (514) 327-9500
Western Canada:
Tel (604) 273-3833

RELATED PRODUCTS
A.F.M. CARPET ADHESIVE, p. 119
A.F.M. POLYURASEAL, p. 173
A.F.M. ALL-PURPOSE POLISH AND WAX, p. 191

DESCRIPTION
NATUR: Agglomerate product of cork granules bound with a synthetic adhesive by pressure and temperature. Fully sanded, 4mm thick "all cork" tile ready for adhesive and a protective urethane finishing. Available in 4 patterns and colors.

REVE: Same as Natur but with the protection of 2 PVC sheets (top coat and backing). Available in 12 texture patterns and colors. The Sea/Land/Wind series are silk-screened patterns available in 5 colors .

Fire resistance for all products is well below maximum allowable by U.S. federal agencies for public areas.

MANUFACTURER'S STATEMENT
A state-of-the-art plant uses 100% of the available material, including the continuously vacuumed dust from tile manufacturing, and the vast quantities of byproducts from the main use of the material, which is the manufacturing of bottle stoppers, an important reason for locating close to these plants.

Packing cartons are of recyclable cardboard.

PRODUCTION

SUSTAINABLY ACQUIRED, OR RENEWABLE RESOURCE

IN-PLANT ENERGY EFFICIENCY & RECYCLING

PACKING / SHIPPING

MINIMUM, RECYCLED, RECYCLABLE PACKAGING

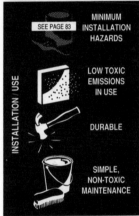

INSTALLATION / USE

SEE PAGE 83

MINIMUM INSTALLATION HAZARDS

LOW TOXIC EMISSIONS IN USE

DURABLE

SIMPLE, NON-TOXIC MAINTENANCE

RESOURCE RECOVERY

OTHER

PRODUCTION

PACKING / SHIPPING

SUSTAINABLY
ACQUIRED, OR
RENEWABLE
RESOURCE

INSTALLATION / USE

SEE PAGE 83

MINIMUM
INSTALLATION
HAZARDS

LOW TOXIC
EMISSIONS
IN USE

DURABLE

SIMPLE,
NONTOXIC
MAINTENANCE

RESOURCE RECOVERY

OTHER

RESEARCH &
EDUCATION
PROGRAMS

PRODUCT
Natural linoleum sheeting

MANUFACTURER
Nairn Floors International Ltd. (Scotland)
560 Weber St. N.
Waterloo, ON N2L 5C6
Tel (519) 884-2602   Fax (519) 888-6548

RELATED PRODUCTS
NAIRNBOND ADHESIVE: Water dispersion adhesive for gluing the flooring.
NAIRNBOND COVE BASE ADHESIVE: Water dispersion adhesive for gluing the seams.

DESCRIPTION
PLAIN LINOLEUM: Sheet goods in 6 solid colors. Available in rolls, width 72" (1.83m), length 36'–92' (11–27.5 lin.m). Jute burlap backing. Standard thicknesses: ⅛" (2.5mm), ³⁄₁₆" (3.2mm), and ¼" (4.5mm). Available in select colors only.

ARMOURFLEX: Marbled linoleum sheet goods, in 24 colors. Available in rolls, width 79" (2m), length 36'–92' (11–27.4 lin.m) and 24" × 24" tile. Jute burlap backing on roll goods, polypropylene backing on tiles. Standard thicknesses: ³⁄₃₂" (2mm), ⅛" (2.5mm), and ³⁄₁₆" (3.2mm). Meets Class 1 flammability requirements.

12" × 12" tiles may be available for the above products. Contact the company for further information. The backing on the tile consists of polypropylene, which makes up 2% of the overall product contents.

MANUFACTURER'S STATEMENT
Nairn has worked with the Saskatchewan government in developing the production of linseed oil for flooring products. Saskatchewan is a major producer of flax. (Argentina is the only other major producer.) Nairn has applied for certification by Environmental Choice's Ecologo, a program established by Environment Canada. A 10-year warranty is given for Nairn linoleum.

MANUFACTURER

Forbo Krommenie BV (Holland)
Imported by
Forbo North America
P.O. Box 32155
Richmond, VA 23294
Tel (804) 747-3714

Western Canada:
Tel (604) 525-4142
Fax (604) 525-3771
Eastern Canada:
Tel (416) 745-4200
Fax (416) 745-4211

RELATED PRODUCTS

Adhesives, pp. 105–6, 116–23
Oil finishes, pp. 184–86

DESCRIPTION

KROMMENIE PLAIN LINOLEUM: Unicolor linoleum sheet goods, in 16 colors. Homogeneous wear surface, with jute backing. Available in rolls 79" wide × 105′ long (width 2m, length 32 lin.m). Standard thicknesses: ³⁄₃₂" (2mm) for domestic use, and ⅛" (3.2mm) for high traffic/commercial use. Seams are heat-welded with matching Linoweld welding rod.

MARMOLEUM: Marbellized linoleum sheet goods, in 68 colors and combinations. Homogeneous wear surface, with jute backing. Sizes and thicknesses similar to plain linoleum, above. Seams are heat-welded with matching Marmoweld welding rod. Marmoform covered skirtings and corners are available.

KROMMENIE CORK: A specially formulated, highly elastic type of unicolored linoleum, consisting of oxidized linseed oil and resins, carefully mixed with ground cork and pigments and calendered onto a jute fabric. Available in rolls (width 2m, length 30–32 lin.m). Standard thicknesses: ¼" (6mm) or ³⁄₁₆" (4.5mm) for commercial use, and ⅛" (3.2mm) for domestic use. Appropriate for sports halls, gymnasiums, tennis halls, physiotherapy and physical rehabilitation centers, as well as bedrooms and nurseries.

KROMMENIE CORKMENT: The basic ingredients for Corkment are identical to Cork, only the cork particles are coarser, and no pigments are added. It is an ideal underlay to increase resilience, and to act as an impact, sound, and thermal insulation.

MANUFACTURER'S STATEMENT

Krommenie linoleum has been in production since 1899. The material is still made free of chemicals found to be hazardous, and made from natural ingredients. Forbo's advanced calendering technology and the ability to impregnate pigment into extremely fine particles of wood flour produces bright colors that never lose their sheen.

Forbo is working together with an association recycling PVC in Germany for waste from their synthetic products. The mill waste from linoleum production is fed back into the manufacturing process, minimizing scrap disposal.

PRODUCTION

SUSTAINABLY ACQUIRED, OR RENEWABLE RESOURCE

IN-PLANT ENERGY EFFICIENCY & RECYCLING

LOW EMISSIONS PLANT

PACKING / SHIPPING

INSTALLATION / USE

SEE PAGE 83

MINIMUM INSTALLATION HAZARDS

LOW TOXIC EMISSIONS IN USE

DURABLE

SIMPLE, NONTOXIC MAINTENANCE

RESOURCE RECOVERY

OTHER

RECYCLED
CONTENT

PRODUCTION

PACKING / SHIPPING

SEE PAGE 83

MINIMUM
INSTALLATION
HAZARDS

DURABLE

SIMPLE,
NON TOXIC
MAINTENANCE

INSTALLATION / USE

RESOURCE RECOVERY

OTHER

PRODUCT
Resilient rubber flooring products with 100% recycled rubber content

MANUFACTURER

Carlisle Tire & Rubber
1415 Ritner Highway, Box 99
Carlisle, PA 17013
Tel (717) 249-1000

DESCRIPTION

PLAYGUARD: Resilient playground surface for use under playground equipment. Tile cast from recycled rubber granules and binders. A dye provides color throughout the thickness of the material. Tile size 2' × 2' × 2" or 3¼" thick. Available in 3 colors: rust, gray, black. Playguard FR 1 Series Tile has Class A flammability rating for use as a roofing material.

PLAYGUARD SUPERTOP: Resilient playground surface for use under playground equipment. Tile cast from recycled rubber granules and binders. Has special top layer for superior color stability, scrubbability and stain resistance. Tile size 2' × 2' × 2" or 3¼" thick. Available in 8 different color blends.

SOFTPAVE: Tile cast from recycled rubber granules and binders. A dye provides color throughout the thickness of the material. Tile size 2' × 2' × 1" thick. Available in 5 colors: rust, gray, black, red flecked, and green flecked. Softpave FR 1 Series Tile has Class A flammability rating for use as a roofing material.

SOFTPAVE SUPERTOP: Tile cast from recycled rubber granules and binders. A thermoplastic top layer is designed especially for pool and spa decks. Tile size 2' × 2' × 1" thick. Available in 4 colors: blue/white, aqua/gray, mocha, and granite.

MANUFACTURER'S STATEMENT

The recycled rubber products were developed as a means of reducing solid waste from the company's tire manufacturing plant and making it marketable. Currently, this accounts for less than 50% of the required material for production of these products. The balance is purchased on the open market from recycled rubber brokers.

PRODUCT
Rubber tile with 100% recycled rubber content

MANUFACTURER

Dodge-Regupol Incorporated
P.O. Box 989,
Laurel & Manor Streets
Lancaster, PA 17603
Tel (717) 295-3400, (800) 322-1923
Fax (717) 295-3414

Phoenix Floor and Wall Products
111 Westmore Dr.
Rexdale, ON M9V 3Y6
Tel (416) 745-4200
Fax (416) 745-4211

RELATED PRODUCTS

SYNTHETIC SURFACES #78H: 2-part solvent-free epoxy adhesive.
XL CORPORATION RUBBER BOND: 1-part solvent-free latex adhesive.

DESCRIPTION

EVERLAST: Nonlaminated, 1-piece floor tile consisting of polymerically (urethane) bound recycled rubber mixed with colored ethylene propylene diene monomer (EPDM) granules or pigmented styrene-butadiene rubber (SBR).

Everlast is antiskid and spike resistant. Suitable for use in weight rooms, golf pro shop areas, aerobic/fitness centers, walkways and corridors, ice rinks, and locker rooms. It will endure over ten years of heavy usage. Standard sizes: $18'' \times 18'' \times \frac{1}{4}''$ and $36'' \times 36'' \times \frac{3}{8}''$. Rolls up to 36" wide, and custom sizes available on special orders, with minimum quantities required.

Available in 6 solid colors (gray, green, red, brown, white, blue) and 12 standard color combinations. Custom blended colors available.

MANUFACTURER'S STATEMENT

Formed in 1989 as a joint venture between Dodge Cork Co. Inc. and Berleburger Schaumstoffwerk GmbH (Germany). State-of-the-art German technology is two generations ahead of anything in North America.

Every 5 square feet of Everlast rids the earth of approximately one scrap tire.

Other products made by Dodge-Regupol out of recycled rubber: molded playground tiles, recreational molded forms, handicapped access ramps, interlocking rubber pavers, landscape products, molded industrial floors, load securement devices, steel coil holders, rubber curbing for playgrounds.

PRODUCTION

RECYCLED CONTENT

IN-PLANT ENERGY EFFICIENCY & RECYCLING

PACKING / SHIPPING

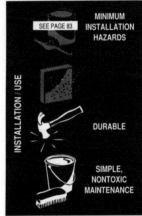

INSTALLATION / USE

SEE PAGE 83

MINIMUM INSTALLATION HAZARDS

DURABLE

SIMPLE, NONTOXIC MAINTENANCE

RESOURCE RECOVERY

OTHER

RECYCLED CONTENT

PRODUCTION

PACKING / SHIPPING

SEE PAGE 83

MINIMUM INSTALLATION HAZARDS

INSTALLATION / USE

DURABLE

SIMPLE, NONTOXIC MAINTENANCE

RESOURCE RECOVERY

OTHER

PRODUCT
Resilient cork/rubber flooring products with 100% recycled rubber content and cork waste byproduct

MANUFACTURER
Global Technology Systems, Badger Cork Division
26112 110th St., P.O. Box 25
Trevor, WI 53179
Tel (414) 862-2311   Fax (414) 862-2500

DESCRIPTION
COMPOSITION CORK ROLLS: Cork granules are combined with a polymeric binder, for commercial grade products. A broad range of materials having a variety of uses, from bulletin boards to flooring and gaskets. Products are used in vibration and thermal insulation, and sound deadening. Standard roll sizes are 37" and 50" wide, with custom widths upon request.

MLA is a low density, medium granule cork used as underlay, and for sound deadening, not for use as a facing material. *MLX* and *ML* are medium granule, intermediate density, with a good appearance, for bulletin boards and other decorative uses. *FI* is a fine granule, intermediate density, high quality, general purpose cork, generally used for decorative applications. *FL* is a lower quality, fine granule cork material, suitable for underlay, sound deadening, or inexpensive bulletin board applications.

RUBBER/CORK SHEETS: Traditionally used for sealing and gasketing, more recent innovations include compressive pads and matting. Sheets are sliced from standard size blocks, and standard thicknesses range from .031" to .250". Custom thicknesses are available.

SP-52 is used as cushioning for furniture, and insulation on windows and doors. Consists of neoprene sponge and cork granules that are moderately oil resistant. *SP-74* is excellent for noise reduction, matting, cushioning, fatigue and compression mats under equipment and furniture. Consists of SBR sponge and cork granules and is highly compressible.

RUBBER MATERIALS: Generally used as walk pads, barrier pads, or matting, with recent developments producing matting for decorative applications: tennis courts, marine decking, and skidproof pool decking. Granules are synthetic rubber in a polymeric binder. Rolls come in standard widths of 32", 36", 42", and 48".

GT-403 is for use in the sport and athletic surfacing. *GT-430* is for use in the playground pad market. *GT-406* has a high density, and good wear properties, for use in high traffic areas.

MANUFACTURER'S STATEMENT
The cork and cork/rubber products can be manufactured with a protein binder or a natural rubber, for a fully natural material. The cork in the composite materials acts as a flame retardant by reducing the toxic fumes of synthetics in the case of fire.

The rubber, cork, and composite materials are resource efficient, and are examples of uses of reclaimed and salvaged materials. The cork is obtained from the cork bottle stopper industry, and Global Technology Systems owns and operates a state-of-the-art plant in Portugal. The rubber roll products utilize granulated rubber buffings from ground tires.

Seamless, hard-surface flooring system

MANUFACTURER

General Polymers
145 Caldwell Dr.
Cincinnati, OH 45216
Tel (513) 761-0011
Fax (513) 761-4496

General Polymers Canada
405 The West Mall, Suite 700
Etobicoke, ON M9C 5J1
Tel (416) 621-3988

DESCRIPTION

MACROSEPTIC®: Seamless, hard surface flooring system, with Intersept® to inhibit the growth of harmful bacteria and fungi. Suited for working environments that demand high performance and hygienic freshness: research labs, hospital operating rooms, food processing plants, grocery stores, kitchens, locker rooms, showers, restrooms, cafeterias, animal housing, and manufacturing.

Heavy duty flooring: can be applied as a ¼" to ⅜" thick protective surfacing on new concrete, or as a resurfacing on worn or deteriorated substrates.

Moderate wear flooring: can be applied in a slurry system at ¹⁄₁₆" to ⅛" thickness. Available in smooth or nonskid surface.

Thinset terrazzo systems: limitless designs of patterns and color.

Each system is available in a variety of colors and quartz aggregate combinations (encapsulated in a clear epoxy resin).

MANUFACTURER'S STATEMENT

General Polymers has been developing composition resin flooring and coating systems to solve the challenges presented by specialized environments. Chemists and technicians use state-of-the-art polymer technology in research and development to meet the broadest array of flooring needs.

General Polymers is a partner of the Envirosense™ consortium who provide products manufactured with Intersept®, a low toxicity, broad spectrum biocide.

NOTE

The addition of biocides to materials may not be an appropriate practice for many applications. Biocides are not a substitute for adequate cleaning and maintenance.

PRODUCTION

IN-PLANT
ENERGY
EFFICIENCY
& RECYCLING

PACKING / SHIPPING

INSTALLATION USE

LOW TOXIC
EMISSIONS
IN USE

DURABLE

SIMPLE,
NONTOXIC
MAINTENANCE

RESOURCE RECOVERY

OTHER

RESEARCH &
EDUCATION
PROGRAMS

PRODUCTION

IN-PLANT ENERGY EFFICIENCY & RECYCLING

LOW EMISSIONS PLANT

PACKING / SHIPPING

MINIMUM, RECYCLED, RECYCLABLE PACKAGING

INSTALLATION / USE

SEE PAGE 83 — MINIMUM INSTALLATION HAZARDS

LOW TOXIC EMISSIONS IN USE

DURABLE

SIMPLE, NONTOXIC MAINTENANCE

RESOURCE RECOVERY

REUSABLE, SALVAGEABLE

OTHER

PRODUCT
Commercial vinyl sheet flooring

MANUFACTURER
Tarkett AB (Subsidiary of STORA Group, Sweden)
Imported by

Tarkett Inc.
800 Lanidex Plaza
Parsippany, NJ 07054
Tel (201) 428-9000

Tarkett Inc.
215 Carlingview Dr., Suite 112
Rexdale, ON M9W 5X8
Tel (416) 675-1133
Fax (416) 674-1146

RELATED PRODUCTS
Adhesives, pp. 105–6, 116–23

DESCRIPTION
EPOC, EPOC AKUSTIK, EPOC UNIC, TARKETT EXTRA, AND TARKETT STANDARD PLUS: Composite vinyl floorings made of a glass fiber carrier, PVC, and mineral fillers.

Epoc series available in 9 colors, Tarkett Extra and Tarkett Standard Plus available in 7 colors.

EMINENT, OPTIMA AND GRANIT: Homogeneous vinyl floorings made of PVC and mineral fillers.

Eminent (2mm and foam/foil) available in 5 colors, Optima (2mm and foam/foil) and Granit (2mm and foam/foil) available in 9 colors.

MANUFACTURER'S STATEMENT
Tarkett AB has developed vinyl flooring products for looselay, eliminating the need for adhesives, and making removal and recycling easier. The company is a member of a European PVC manufacturers' association and PVC recycling group.

Filter technology for collecting lost plasticizer material during production reduces emissions by 10%–16%.

All packing materials are marked with the German designation for recyclable materials—paper and polyethylene. Recycling programs for in-office paper are established, and in-plant recycling has been practiced for over 15 years.

INSTALLATION MATERIALS
• CAULKING • JOINT COMPOUNDS •
• MORTAR • GROUT AND ADHESIVES •

CAULKING

Caulking is used to fill a gap or create a seal where some flexibility is required. Traditional caulking materials were linseed oil putties, tree resins, and asphalt. These were messy to handle, and became brittle over time. Modern caulkings for construction are made primarily from synthetic polymers, some of which are from the same chemical families as paints—latex, acrylics, and urethanes. Others, such as those made from silicones, polychloroprenes, polysulfides, and butyls, are quite different from paint bases. The most common caulkings for indoor use are made from acrylic latex and silicones. Most other caulkings are high performance formulations for outdoor use, where weather resistance is important.

Because caulkings are used in small quantities in most interior construction, their environmental impact and toxicity are less important overall than other materials. However, some caulkings are moderately toxic to handle and will offgas for several weeks. These include the solvent-based varieties like urethanes, polychloroprenes (neoprene), polysulfides, and butyls. This type of caulking is formulated with hazardous solvents, such as acetone, methyl ethyl ketone, toluene, xylene, and alcohols.

When these caulkings are applied, a portion of the solvent, along with other gases from curing, such as pentanes, hexanes, octanes, and benzenes, are released quickly until a "skin" forms on the material. The remaining solvent and gases are released much more slowly and can take months to disappear. These gases are essentially irritants in low concentrations, but some, such as benzene, are carcinogenic. Most of the more hazardous caulkings are intended for outdoor use.

When large quantities of caulking are used indoors for sealing vapor barriers and caulking air leaks in insulated walls and ceilings, selection of a safer material becomes much more important. Quantities of urethane, polysulfide, butyl, and polychloroprene caulkings should be avoided indoors. These can all be identified both by labels and by the strong sulfurous or solvent odors they emit.

The environmental impact of manufacturing a caulking is similar to that for a paint of the same type. Those which are water based and contain little petroleum solvent and no chlorinated materials are the most benign to manufacture. Those which contain solvents, such as polychloroprene, entail more environmental risk and worker exposure in manufacturing.

SAFER CONVENTIONAL CAULKINGS

LATEX CAULKING

Latex caulking is made with synthetic latex (styrene-butadiene rubber) mixed with fillers, glycols, and colorants. It is compatible with paints, and is similar in environmental impact and toxicity risk to latex paints. It can be cleaned up with water when fresh. Though latex caulkings may also contain fungus-retarding biocides, exposure to them is considered a lesser health risk than to the biocides in paints, due to the smaller quantities of caulkings used. Latex caulkings are the safest of the paintable varieties.

ACRYLIC CAULKING

Acrylic caulking contains acrylic resins (methacrylates), usually blended with latex. It also contains glycols, petroleum solvents and fungicides. It has similar air pollution and chemical risk profile to acrylic-latex paints, and is slightly more hazardous than latex alone, due to its additional solvents and resins. Acrylic caulking can be cleaned up with water when fresh.

PLAIN SILICONE CAULKING

Plain silicone caulking contains silicone resin (siloxanes) and acetic acid (vinegar). It is produced by a process requiring chlorinated solvents, resulting in air and water pollution, as well as toxic waste problems. The advantage of silicone is that it "vulcanizes" when setting, that is, it forms a chemically stable rubber which emits very little odor, is resistant to humidity and heat, and is very durable. Once cured, it has the fewest health risks of any available indoor caulking. The acetic acid does produce a strong vinegar odor while curing, which usually lasts a few hours. The odor is unpleasant but not harmful.

Silicone caulking must be cleaned up with solvents. It cannot be painted, and does not adhere well to rough wood or concrete. The main disadvantage of plain silicone is that it supports fungus growth, and will become stained within a few months if used in a humid location.

SILICONE BATH CAULKING

Silicone bath caulking is similar to plain silicone, but it contains toxic fungicide additives and is, therefore, more hazardous to produce and handle. It is used where persistent moisture occurs, such as in bath enclosures, and should be limited to these uses.

LOW TOXICITY AND ALL NATURAL CAULKINGS

Specialty suppliers also make low toxicity caulkings. These are formulated with synthetic resins, but contain little or no hazardous solvents or fungicides. These may not last as long or be as flexible as conventional caulkings, but they are quite adequate for most interior uses, such as draft sealing and setting windows and doors. They are not recommended for humid locations or exterior use.

JOINT, PATCHING, AND CEILING TEXTURING COMPOUNDS

Joint and patching compounds are composed primarily of a gypsum base, but also contain other minerals such as mica, talc, limestone, and clay as fillers, and to improve the compounds' workability and resistance to shrinkage. They also contain adhesives, usually polyvinyl alcohol and polyvinyl acetate, preservatives (mercury compounds and other toxins), and antifreeze in the premixed varieties. Ceiling texturing compounds are similar, but contain perlite or polystyrene as an aggregate.

All joint fillers are mild caustics and will dry the skin on contact. They also require sanding, which is a very dusty operation, exposing tradespeople to irritating dusts. The work area should be separated from any occupied areas during sanding and everyone in the work area should wear an approved respirator mask. Though the dust from joint fillers is not dangerous unless it is inhaled daily, asbestos was a common ingredient until the 1970s, and may still be found in older buildings. Dust and flakes from old fillers are dangerous and should be treated as if they contain asbestos. (Check with local health authorities or the nearest Environmental Protection Agency (EPA) office for more information.)

PREMIXED COMPOUNDS

The premixed varieties contain gypsum which has not been roasted fully to remove its chemically bound water. They will set only by air drying, while fast-setting fillers set by reacting with water. They usually contain vinyls like polyvinyl acetate, stabilizers to keep them from becoming lumpy, and antifreeze to make them frost resistant. Most also contain biocides to prevent spoilage. As with water-based paints, the biocides are the most hazardous ingredient.

Because they are over half water by weight, a high transportation cost must be paid for the small convenience of a ready mixed product. They are packaged in both nonrecyclable plastic and a cardboard box.

DRY-MIXED AND FAST-SETTING COMPOUNDS

Dry-mixed varieties are packaged in paper bags which usually have a plastic coating, making the bags nonrecyclable, but they are much lighter to ship than premixed types. The composition of the standard-setting types is similar to that of the premix above, except that it is less likely to contain stabilizers, biocides, and antifreeze because they are not necessary in dry mixes.

The fast-setting type contains primarily calcined gypsum, which has been heated to remove the water, and will set quickly when water is added again. Though it takes more energy to make, it may contain fewer additives than the premixed or dry-mixed types, making it safer to manufacture and handle.

LOW TOXICITY JOINT FILLERS

A few of the low toxicity paint manufacturers make joint fillers which are carefully formulated without biocides, and with a minimum content of adhesives and stabilizers. They are usually dry-mixed types, as these manufacturers recognize the energy waste of shipping a premixed product. They may be slightly more difficult to handle than conventional types, but are a more appropriate product for people with sensitivities. They are also gypsum based.

MORTAR

True mortar was used traditionally for setting ceramic tile on both floors and walls. It is made from simply cement, lime, and sand but requires rare skill to apply, especially on walls. Very little tile setting is now executed this way. Most is applied with either a plastic resin adhesive (see SPECIALIZED ADHESIVES, below), or with "thinset mortars." Resin adhesive is still prefered for wall tile work in locations with light use because it is easy to use, but it is not appropriate for floor tile. Thinset mortars are made from modified cements and sand, but are easier to work with than true mortar. They are safer to handle and more reliable for tile setting than resin adhesive.

Thinset mortars are sold in a dry-mixed form, which is simply mixed with water for setting tile on concrete and concrete block bases. For use on wood bases, an acrylic additive improves the mortar's adherence. The acrylic is water-based, has no odor, and is safe to handle. Mortars also cause less environmental impact in their production than adhesives.

GROUT

There are several types of grout for tile setting, ranging from traditional cement mortar type (sanded grout) to high performance, epoxy- or silicone-modified grouts. Plain cement grouts are good choices for floor tile. They do not support fungus growth, but are not resistant to acids (such as fruit juice or vinegar) and must be sealed to reduce porosity. More advanced cement grouts contain an additive which makes them acid-resistant and very durable. These are the best for floor tile in high use areas. All grouts in kitchens or bathrooms should be sealed with a water-resistant sealer. (See the FINISHES section.)

Basic wall tile grouts may contain gypsum, methyl cellulose, and other agents to achieve a fine, workable grout which sets quickly. Unfortunately, they are prone to staining, and support the growth of fungus. Once stained, only toxic cleaners or bleaches can restore them. When wall tile grout is sealed, it is easier to maintain. Using a safe mildew retardant when cleaning, such as one containing borax and boric acid, will help control fungus.

Epoxy-, acrylic-, or silicone-modified grouts are much more chemically complex, and introduce more environmental impact and risks. These are more appropriate for heavy-use situations, as they are more durable and require lower maintenance, but are not necessary for residential or light commercial installations.

GENERAL PURPOSE ADHESIVES

Adhesives, like paints, are either solvent based or water based. Solvent-based adhesives contain synthetic resins, and xylene, toluene, 1,1,1 trichloro-ethane, acetone, or other solvents which are environmental hazards and toxic to handle. Water-based, latex adhesives like latex flooring cement or contact cement are safer to handle because they contain smaller quantities of solvents, and can be cleaned up with water. Plain white glue is a low toxicity, water soluble, casein- (milk protein) or polyvinyl acetate-based product, which is safer than latex or solvent adhesives. It is suitable for woods, paper, and fabric, but it is not moisture resistant.

Both the solvent used to clean up solvent-based adhesives, as well as the unused adhesive itself, are hazardous wastes and should be taken to a hazardous waste collection service.

CONTACT CEMENTS

Contact cements contain synthetic resins or natural latex, either in a water base or petroleum solvent. The solvent-based type normally contains xylene, toluene, formaldehyde, hexane, naphthalene, phenol, or petroleum spirits. The water-based type usually contains ammonia, ethanol, and glycols, as well as preservatives to inhibit microorganisms. Water-based contact cement is the safest choice for laminate installation. It performs well and can be cleaned up with water. The solvent-based variety has no special advantage and should be avoided, as it is toxic and highly flammable.

CONSTRUCTION ADHESIVES

These are similar to solvent-based contact cements, except that they are thicker and can be applied with a caulking gun. They contain synthetic rubber, solvents, tackifier-resin, which makes them very sticky, and filler to make them thicker. Depending on the type, they may also contain urethane (isocyanates), and other plastics. Large amounts of construction adhesive are sometimes used to glue subfloors down, and to apply rigid insulation and wallboard. This leaves a large quantity of glue inside the building which will offgas for a long time, and it creates a good deal of waste from the empty glue tubes. For most applications, such as installing subfloors, the use of glue can be avoided entirely by using a screw gun.

SPECIALIZED ADHESIVES

EPOXY GLUES

Epoxy glues are made with epichlorohydrin, phenols, and glycerols. The two-part variety (consisting of a resin and a hardener) will produce chemical burns on contact with skin, and irritation of the respiratory system and eyes from vapor inhalation. Epoxy manufacturing is an environmentally hazardous industry with emission and toxic waste problems. Handling of the raw materials is linked to skin and lung cancer in industrial workers, and the resins are suspected of causing liver and kidney damage. Epoxy is sometimes used for concrete repair, or for very durable finishing for wood and concrete floors. The handling of any large quantity of epoxy should be done only by professionals using appropriate skin protection and respirators, and with adequate ventilation. There are no uses for two-part epoxy in common construction that cannot be met with safer materials. There are, however, safer epoxy formulations, such as tile-setting adhesives and mortar, which do not have a separate hardener and can be cleaned up with water.

RESIN GLUES

Resin glues based on formaldehyde are the most common types used in wood products manufacture. (See INTERIOR CONSTRUCTION PANELS.) They are available as one-part "plastic resin glues" or "resorcinol glues," and as two-part "UF glues" (urea formaldehyde) for woodworking. The two-part glues are more hazardous to handle, and are difficult to mix correctly. Both types are far more hazardous to handle than plain white glues, and there is usually no reason why they should be necessary for interior woodwork. They are high performance glues which should be reserved for special applications like laminating and boatbuilding.

STARCH GLUES (WALLPAPER PASTES)

Starch glues, or wallpaper pastes, contain vegetable starch, and may contain other plant and animal extracts such as casein (milk protein). They also usually contain preservatives because the compound will readily support fungus growth. The safest preservative used in water-based glues is boric acid or borax, which is found in some low toxicity starch glues. They are used primarily for wallpaper hanging, and by artists.

FLOORING, TILE, AND CARPET ADHESIVES

These are similar to contact cements and construction adhesives, and are available in both solvent-based and water-based varieties. The manufacture of the solvent types is more hazardous to the environment and industry workers, and their use involves more toxic risk and hazardous waste than water-based varieties. They are highly flammable and potentially explosive.

LOW TOXICITY FLOORING ADHESIVES

These are specially formulated, water-based adhesives which contain virtually no aromatic solvents, and have almost none of the characteristic strong odor of other adhesives. These are now carried by many flooring suppliers. They are safer for the floorlayer, safer for building occupants, and have less manufacturing impact and toxic waste. There are low toxicity formulations for every type of flooring installation.

DRY ADHESIVES

Factory-applied dry adhesives are a viable alternative for some lines of vinyl tile, wood parquet, and carpet. (See CARPETING, pp. 56–57, and FLOORING, pp. 82–83.) The adhesive is a solvent-free composition, which is very safe to handle and has little odor. Adhesives which are applied under factory-controlled conditions minimize worker exposure and toxic wastes. The main problem with this system is that it generates waste plastic or waxed paper from the protective backing sheets which are removed during installation. One carpet manufacturer using this method is currently setting up facilities to recover and recycle this backing film.

Solvent weld for sheet flooring is a very potent solvent capable of dissolving the edges of vinyl and allowing two sheets to be fused together. It is a hazardous material which must be used with copious ventilation. Though it contributes to regional air pollution, both during manufacture and use, it is only used in small quantities. Heat-welding, or seam-welding, is a safer alternative used for true linoleums as well as for vinyl. (See FLOORING, p. 83.) The method involves melting a colored filler rod, to match the floor, achieving a seamless installation.

PRODUCTION

LOW
EMISSIONS
PLANT

PACKING / SHIPPING

MINIMUM,
RECYCLED,
RECYCLABLE
PACKAGING

INSTALLATION / USE

MINIMUM
INSTALLATION
HAZARDS

LOW TOXIC
EMISSIONS
IN USE

DURABLE

RESOURCE RECOVERY

OTHER

RESEARCH &
EDUCATION
PROGRAMS

PRODUCT
Low toxicity caulking compound

MANUFACTURER

A.F.M. Enterprises Inc.
(American Formulating and Manufacturing)
1140 Stacy Court
Riverside, CA 92507
Tel (714) 781-6860, 781-6861   Fax (714) 781-6892

DESCRIPTION

CAULKING COMPOUND: Flexible, water-based, water-resistant, single component, pigmented rubber emulsion-type caulking compound.

Possesses thixotropic properties that eliminate stringiness and allow ease of application. Increased resin solids extend adhesive bond strength. Protection against corrosion and fungal attack. Water resistant, adheres to numerous substrates, and is stable upon aging.

DYNO FLEX: Water emulsion base system mastic compound and sealer.

Impermeable under most weather conditions. Good sound control barrier (e.g. on metal roofs). May be applied directly over asphalt shingles and roll roofing. It repairs fractured seams and leaks on metal roofs, heating and air conditioning ducts.

Comes in natural, white or gray. Other colors available on special order.

MANUFACTURER'S STATEMENT

Toxicity tested by the Environmental Protection Agency and independent laboratories without the use of animals. Products found widely accepted by those with allergies and chemical sensitivity.

All products are formulated as nonhazardous and stable, and do not contain bactericides or fungicides that are classified as phenol mercury acetates, phenol phenates, or phenol formaldehyde. Contain no aromatic and aliphatic solvents which are present in many other products.

Cans and boxes are made from recycled materials.

PRODUCT
Low toxicity, colored caulking compound

MANUFACTURER
L. M. Scofield Company
6533 Bandini Blvd.
Los Angeles, CA 90040-3182
Tel (213) 723-5285
Fax (213) 722-6029

DESCRIPTION
LITHOCHROME COLORCAULK SEALANTS: Multipart polyurethanes used to caulk and seal joints of concrete, masonry, metal, or wood. Available in *Flow* and *Non-Sag* grades. Color matched to all Scofield standard, special, and custom colors.

MANUFACTURER'S STATEMENT
The vast majority of L. M. Scofield's products are water based and low odor. A company policy has been adopted so that all new product development, and discontinued products which contain more toxic substances (i.e. heavy metals), will be replaced with only low toxicity, water-based formulations.

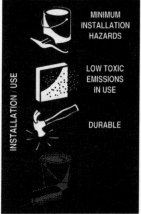

MINIMUM
INSTALLATION
HAZARDS

LOW TOXIC
EMISSIONS
IN USE

DURABLE

PRODUCTION

LOW
EMISSIONS
PLANT

PACKING SHIPPING

MINIMUM,
RECYCLED,
RECYCLABLE
PACKAGING

INSTALLATION / USE

MINIMUM
INSTALLATION
HAZARDS

LOW TOXIC
EMISSIONS
IN USE

RESOURCE RECOVERY

OTHER

RESEARCH &
EDUCATION
PROGRAMS

PRODUCT
Low toxicity joint and patching compound

MANUFACTURER
A.F.M. Enterprises Inc.
(American Formulating and Manufacturing)
1140 Stacy Court
Riverside, CA 92507
Tel (714) 781-6860, 781-6861   Fax (714) 781-6892

DESCRIPTION
JOINT COMPOUND: All-purpose, nonasbestos joint and patching compound. Low shrinkage, nonflammable, nonhazardous, low odor. Does not require the use of toxic additives.

MANUFACTURER'S STATEMENT
See p. 108.

PRODUCT
Low toxicity, all-purpose joint cement and texture compound

MANUFACTURER

Murco Wall Products, Inc.
300 N.E. 21st St.
Fort Worth, TX 76106-528
Tel (817) 626-1987   Fax (817) 626-0821

DESCRIPTION

M-100 HYPO JOINT COMPOUND: A starch-type, powdered, all-purpose joint cement and texture compound formulated with inert fillers and natural binders only. It does not contain any preservatives or slow releasing compounds. It mixes easily with tap water.

Conventionally drying product, smooth application, and ample working time. Designed for embedding tape, fill coats, and finishing over drywall joints, corner metal, trims, and fasteners. It offers controlled shrinkage to reduce edge cracking.

Also used for textured finishes.

MANUFACTURER'S STATEMENT

Murco's hypoallergenic products for the chemically sensitive offer non-asbestos formulation, low odor, high finished job quality, few application restrictions, and are cost comparable to conventional paints and joint compounds.

PRODUCTION

SUSTAINABLY ACQUIRED, OR RENEWABLE RESOURCE

PACKING / SHIPPING

INSTALLATION USE

MINIMUM INSTALLATION HAZARDS

LOW TOXIC EMISSIONS IN USE

RESOURCE RECOVERY

OTHER

MINIMUM
INSTALLATION
HAZARDS

LOW TOXIC
EMISSIONS
IN USE

DURABLE

PRODUCTION

PACKING / SHIPPING

INSTALLATION / USE

RESOURCE RECOVERY

OTHER

PRODUCT
Miscellaneous thinset mortar systems

MANUFACTURER

Laticrete International, Inc.
1 Laticrete Park N.
Bethany, CT 06525-3498
Tel (203) 393-0010, (800) 243-4788
Fax (203) 393-1684

Ontario:
Tel (519) 576-9840
Fax (519) 745-3822
Quebec:
Tel (514) 346-0107
Fax (514) 346-4428
Western Canada:
Tel (604) 732-1883
Fax (604) 732-5556

RELATED PRODUCTS

DRYBOND #40 ACRYLIC ADMIX: Acrylic latex liquid for mixing with grouts, thinset mortars, mud beds, leveling mortars, and all Portland cement mixes. Renders mortars and grouts durable and permanent. Especially recommended when installing vitreous tiles like mosaics, monocottura tile, porcelain tile, and slate. Also recommended as a grout additive. Easy to clean with water, nonflammable and low toxic.

DRYBOND #333 FLEXIBLE THINSET AND GROUT FORTIFIER: Flexible additive for use with factory-prepared thinset mortars and ceramic tile grouts. Recommended for use with Drybond Floor 'N Wall Thinset and Laticrete sanded and unsanded grouts. Renders the thinset 6 times more flexible and 30% stronger. Easy to use, and mixes with any thinset or grout. Water cleanable, fast setting, nontoxic and nonflammable liquid.

DESCRIPTION

DRYBOND FLOOR 'N WALL THINSET MORTAR: Versatile dry-set mortar for floor and wall tile installations. Easy to use and water cleanable. Nonflammable and nontoxic during storage, application, and after hardening. For use installing ceramic tile, quarry tile, pavers, mosaics, marble, and natural stone over concrete, masonry, gypsum wallboard, Wonderboard, cement board, cement plaster, and expanded polystyrene insulation.

LATICRETE #4237 LATEX THINSET MORTAR ADDITIVE, LATICRETE #211 CRETE FILLER POWDER: #4237 Additive is a specially designed latex additive to be used with #211 Crete Filler Powder to make a high strength latex, thinset mortar. Suitable for all kinds of ceramic tile and natural stone. #211 Crete Filler Powder is a blend of Portland cement and graded aggregates. Applications include interior and exterior, wet and dry areas, walls, floors, and ceilings. Easy to use, nonflammable, safe to store and mix.

LATICRETE #226 THICKSET MORTAR MIX: Factory-prepared blend of Portland cement and graded aggregates. Designed for use with Laticrete #3701 Latex Mortar Additive to produce a latex Portland cement mortar. For full bed (or conventional) installation of ceramic tile and natural stone. Applications include interior and exterior, wet and dry areas, walls, floors, and ceilings. Easy to use, nonflammable, safe to store and mix.

MANUFACTURER'S STATEMENT
See p. 114.

PRODUCT
Low toxicity tile grout

MANUFACTURER
A.F.M. Enterprises Inc.
(American Formulating and Manufacturing)
1140 Stacy Court
Riverside, CA 92507
Tel (714) 781-6860, 781-6861   Fax (714) 781-6892

DESCRIPTION
TILE GROUT: Ready-to-use emulsion base product formulated in a water-based system. For use as tile grout (on narrow joints only). Not for use on porous tiles.

Nonflammable, nonhazardous, without the need for allergenic admixtures and additives used in conventional dry grout systems.

MANUFACTURER'S STATEMENT
See p. 108.

PRODUCTION

LOW EMISSIONS PLANT

PACKING / SHIPPING

MINIMUM, RECYCLED, RECYCLABLE PACKAGING

INSTALLATION / USE

MINIMUM INSTALLATION HAZARDS

LOW TOXIC EMISSIONS IN USE

RESOURCE RECOVERY

OTHER

RESEARCH & EDUCATION PROGRAMS

LOW TOXIC
EMISSIONS
IN USE

DURABLE

SIMPLE,
NON-TOXIC
MAINTENANCE

PRODUCTION

PACKING / SHIPPING

INSTALLATION / USE

RESOURCE RECOVERY

OTHER

PRODUCT
Low toxicity epoxy grout

MANUFACTURER

Laticrete International, Inc.
1 Laticrete Park N.
Bethany, CT 06525-3498
Tel (203) 393-0010, (800) 243-4788
Fax (203) 393-1684

Ontario:
Tel (519) 576-9840
Fax (519) 745-3822
Quebec:
Tel (514) 346-0107
Fax (514) 346-4428
Western Canada:
Tel (604) 732-1883
Fax (604) 732-5556

DESCRIPTION

LATAPOXY SP-100: Stainless, colorfast epoxy grout for ceramic tile, quarry tile, pavers, floor brick, packing house tile and stone. For commercial use in food service applications, as it will not be stained by most foods, beverages, and cleaning agents. Chemical resistant, durable and easy to maintain. No noxious odor, nonflammable, cold water cleanup.

Available in 3 colors: bright white, sterling silver, and gemstone.

May also be used as an installation material. Will produce high bond strength to surfaces such as plywood, steel, plastic laminate, concrete, ceramic tile, and vinyl.

MANUFACTURER'S STATEMENT

Since its inception in the late 1950s, Laticrete's policy is to produce products that are safe for both personnel, at the time of manufacture, and the consumer, in use. All products are in a water dispersion, and are nonflammable. Materials are screened to be nontoxic (safe for ingestion) and have certification of approval from the FDA.

Waste water is recovered and recycled back into the process.

Even the smaller unit quantities of the products are packed in industrial-type, childproof packaging. The trend to package products in more common, household containers is more economical, but can lead to confusion with recognizable, packaged foodstuffs. Packing containers are made of cardboard or polyethylene, which are from partially recycled materials and are recyclable. Both office and plant participate in paper recycling programs.

PRODUCT
Low toxicity concrete coloring systems

MANUFACTURER
L. M. Scofield Company
6533 Bandini Blvd.
Los Angeles, CA 90040-3182
Tel (213) 723-5285
Fax (213) 722-6029

DESCRIPTION

CHROMIX ADMIXTURES FOR COLOR-CONDITIONED CONCRETE: Pigmented, water-reducing, set-controlling admixtures for colored flatwork and vertical architectural concrete. Colors concrete integrally and increases strength. 20 standard colors. Custom colors by special order.

LITHOCHROME COLOR HARDENER: Ready-to-use, dry-shake, colored or concrete gray hardener for coloring, hardening, and finishing new concrete floors. Streak-free intergrind of pigments, surface-conditioning and dispersing agents, Portland cement combined with hard, graded aggregate. 24 standard colors, concrete gray, and custom colors by special order.

EMERCHROME FLOOR HARDENER: Heavy duty, slip-resistant coating for industrial and commercial flatwork. Concrete gray is stocked, and custom colors by special order.

LITHOCHROME COLORWAX: Colour-matched, heavy duty curing and finishing material for colored concrete flatwork.

MANUFACTURER'S STATEMENT
The vast majority of L. M. Scofield's products are water based and low odor. A company policy has been adopted so that all new product development, and discontinued products which contain more toxic substances (i.e. heavy metals), will be replaced with only low toxicity, water-based formulations.

PRODUCTION

PACKING / SHIPPING

INSTALLATION / USE

MINIMUM INSTALLATION HAZARDS

LOW TOXIC EMISSIONS IN USE

DURABLE

RESOURCE RECOVERY

OTHER

PRODUCTION

SUSTAINABLY ACQUIRED, OR RENEWABLE RESOURCE

LOW EMISSIONS PLANT

PACKING / SHIPPING

INSTALLATION / USE

MINIMUM INSTALLATION HAZARDS

LOW TOXIC EMISSIONS IN USE

DURABLE

RESOURCE RECOVERY

OTHER

RESEARCH & EDUCATION PROGRAMS

PRODUCT
Multipurpose all-natural adhesive

MANUFACTURER
Auro Organic Paints (Germany)
Imported by Sinan Co., Natural Building Materials
P.O. Box 857
Davis, CA 95617-0857
Tel (916) 753-3104

DESCRIPTION
NO. 381-389 NATURAL ADHESIVE: Pure organic binders in water dispersion for gluing cork, tile, linoleum, wall-to-wall carpeting, parquet floors, and wallpaper. Strong adhesive properties, stays permanently elastic, pleasant odor during and after application.

MANUFACTURER'S STATEMENT
Auro products contain no isoparrafinic hydrocarbons (petroleum solvents), which are present in many other products.

Auro has done years of research and development, production and processing of natural substances to arrive at their products. In close cooperation with experienced craftspeople, quality is constantly controlled and improved where possible. Auro was the first producer worldwide to list all ingredients of their environmentally clean products. Plant chemistry's raw materials come from a closed ecological cycle which makes them permanently available.

Auro intentionally maintains small manufacturing plants, producing products for international distribution. This strategy supports local economies and maintains a high quality of product.

Sinan's catalog/newsletter is printed with soybean inks on 100% recycled paper, that has neither been de-inked or bleached.

Miscellaneous all-natural adhesives

MANUFACTURER

Livos Plant Chemistry (Germany)
Imported by

Eco-House (1988) Inc.
P.O. Box 220, Stn. A
Fredericton, NB E3B 4Y9
Tel (506) 366-3529
Fax (506) 366-3577

Livos Plant Chemistry / USA
1365 Rufina Cir.
Santa Fe, NM 97214
Tel (505) 438-3448

DESCRIPTION

#510 LINAMI CORK GLUE: Natural latex base, for cork flooring and wall covering. May also be used for gluing paper and cardboard.

#511 MELINO CARPET AND LINOLEUM GLUE: Natural resin glue is permanently elastic, wheelchair resistant and temperature resistant. For installation of carpets with jute or soft rubber backings, and linoleum (up to 3mm thickness).

#540 LAVO WALLPAPER GLUE: Old-fashioned wallpaper paste, made of pure methyl cellulose, without fungicide or insecticide additives. For use with all types of wallpaper.

MANUFACTURER'S STATEMENT

Livos products are designed to improve the quality of your personal environment. All ingredients used are selected to meet the following criteria: 1) no health threatening materials; 2) ecological soundness of raw material production, manufacturing process, application, and disposal; 3) highest technical quality.

Livos products contain only natural and nontoxic ingredients, providing a pleasant fragrance during application and eliminating continued evaporation of fumes from the dried product. Finishes are harmless to humans, plants, soil, and water. Ingredients are known, natural and renewable, and require no testing on animals.

Livos maintains small manufacturing plants, producing products for local and international distribution. This strategy supports local economies and maintains a high quality of product.

Process acids are recycled and reused. Containers are of recyclable metal or cardboard/paper from recycled materials.

PRODUCTION

SUSTAINABLY ACQUIRED, OR RENEWABLE RESOURCE

LOW EMISSIONS PLANT

PACKING / SHIPPING

MINIMUM, RECYCLED, RECYCLABLE PACKAGING

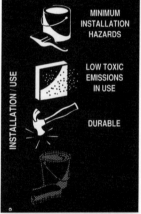

INSTALLATION / USE

MINIMUM INSTALLATION HAZARDS

LOW TOXIC EMISSIONS IN USE

DURABLE

RESOURCE RECOVERY

OTHER

RESEARCH & EDUCATION PROGRAMS

INSTALLATION
Adhesive

PRODUCTION

SUSTAINABLY ACQUIRED, OR RENEWABLE RESOURCE

PACKING / SHIPPING

INSTALLATION / USE

MINIMUM INSTALLATION HAZARDS

LOW TOXIC EMISSIONS IN USE

DURABLE

RESOURCE RECOVERY

OTHER

FAIR BUSINESS PRACTICES

PRODUCT
Miscellaneous all-natural adhesives

MANUFACTURER

Naturhaus (Germany)
Imported by
Lindquist Marketing
115 Garfield St.
Sumas, WA 98295
Tel (604) 826-1547
Fax (206) 988-2207

Lindquist Marketing
P.O. Box 3542, 33118 Whidden Ave.
Mission, BC V2V 2T2
Tel (604) 826-1547

DESCRIPTION

WALLPAPER PASTE: White odor-free powder, easy to dissolve in cold water. PH neutral, free of chemicals. Suited for all wallpaper products (light or heavy materials on absorbing surfaces). Contains methyl cellulose.

#09020 GLUE FOR LINO, CORK, AND CARPET: Good bonding ability, elastic, and waterproof. For linoleum, carpet, cork flooring, and insulation. Contains casein, latex, dammar resin, water, ammonia, and talcum.

MANUFACTURER'S STATEMENT

Naturhaus has an overall philosophy of making available natural, nontoxic products for healthier living. In Germany, their retail outlet sells cork flooring, nontoxic, natural/organic clothing, finishing products, etc.

The finishing products are all natural, and the nontoxic mineral ingredients provide more durable, stable products in humid climates. Traditional ingredients are time-tested, so that toxicity testing is not required. Natural material ingredients also support developing Third World countries by providing a demand for traditional, low environmental impact, sustainable industries and products.

PRODUCT
Miscellaneous low toxicity adhesives

MANUFACTURER
A.F.M. Enterprises Inc.
(American Formulating and Manufacturing)
1140 Stacy Court
Riverside, CA 92507
Tel (714) 781-6860, 781-6861   Fax (714) 781-6892

DESCRIPTION
CARPET ADHESIVE: Low toxicity, water-based acrylic vinyl adhesive (elastometric water-based emulsion). For installation of textiles and most types of carpet. May be used as a multipurpose adhesive.

3-IN-1 ADHESIVE: High bonding strength adhesive recommended for ceramic, vinyl, parquet tiles, and plastic laminate. Elastomeric water-based emulsion type. Comes ready to use, in a creamy, easy-to-spread consistency.

WALLPAPER ADHESIVE: Water-based system adhesive for wallpaper, vinyl wall coverings, etc. Synthetic resin emulsion type. Contains no petroleum solvent components, has increased resin solids that extend the adhesive's strength.

Wall sizing available, on special order.

MANUFACTURER'S STATEMENT
See p. 108.

INSTALLATION
Adhesive

PRODUCTION

LOW EMISSIONS PLANT

PACKING / SHIPPING

MINIMUM, RECYCLED, RECYCLABLE PACKAGING

MINIMUM INSTALLATION HAZARDS

LOW TOXIC EMISSIONS IN USE

INSTALLATION / USE

RESOURCE RECOVERY

OTHER

RESEARCH & EDUCATION PROGRAMS

MINIMUM
INSTALLATION
HAZARDS

LOW TOXIC
EMISSIONS
IN USE

RESEARCH &
EDUCATION
PROGRAMS

PRODUCT
Miscellaneous low toxicity adhesives

MANUFACTURER

Roberts
600 North Baldwin Park Blvd.
City of Industry, CA 91746
Tel (818) 369-7311
Fax (818) 369-7318

Roberts Company Canada Ltd.
2070 Steeles Ave.
Bramalea, ON L6T 1A7
Tel (416) 791-4444
Fax (416) 791-1998

DESCRIPTION

Roberts Earthbond line of products are solvent-free adhesives containing no alcohol, methanol, glycol, or ammonia, hence no associated hazardous vapours and minimal potential for pollution of the environment.

EARTHBOND 6900 MULTI-PURPOSE ADHESIVE: Synthetic latex, low toxicity adhesive designed to adhere inorganic mineral fiber and paper felt-backed vinyl sheet goods, linoleum, sponge back, high density latex foam-backed, needle-punched, latex unitary, woven jute, urethane (only) and polypropylene secondary-backed carpets. May be used over felt, plywood of underlayment quality, terrazo, marble, sound resilient floors which are wax-free, and concrete above, on, or below grade. Porous surfaces should be sealed with shellac before installation.

EARTHBOND 7000 MULTI-PURPOSE ADHESIVE: Synthetic latex, low toxicity adhesive designed to adhere to paper felt-backed vinyl sheet goods, linoleum, rubber tile, sponge-backed, high density latex foam-backed, needle-punched, latex unitary, woven jute, urethane-backed, and polypropylene secondary-backed carpets. May be used over felt, plywood of underlayment quality, terrazo, marble, sound resilient floors which are wax-free, and concrete above, on, or below grade.

EARTHBOND 7100 SHEETGOODS ADHESIVE, EARTHBOND 7600 ALL VINYL FLOORING ADHESIVE: Synthetic latex, low toxicity adhesives designed to adhere all mineral/paper/felt-backed vinyl sheet goods. It may be used over felt, plywood of underlayment quality, existing resilient floors, or concrete above, on, or below grade.

EARTHBOND 7200 COVE BASE ADHESIVE: Low toxicity acrylic adhesive designed to adhere vinyl and rubber cove bases over clean, dry, and structurally sound wall surfaces. Not for use over nonporous surfaces like vinyl wall coverings.

EARTHBOND 7500 CARPET TILE ADHESIVE: Synthetic latex, low toxicity adhesive designed to adhere to all types of carpet tiles, including vinyl-backed and fusion-bonded tiles, and vinyl composition tiles. May be used over felt, plywood of underlayment quality, metal, terrazo, marble, sound resilient floors which are wax-free, and concrete above, on, or below grade. Avoid installations over existing resilient floors that are deeply embossed or have cushion backs. Do not use on wood floors that are treated with fire retardant. Porous and filled surfaces should be sealed with shellac or liquid latex sealers.

EARTHBOND 7700 CONTACT CEMENT: Nonflammable, solvent-free emulsion of vinyl polymers. Versatile contact adhesive that provides strong, heat-resistant bonds and covers up to 2½ times more than solvent-based contact cement. For bonding plastic laminates, plywood panels, wallboard, hardboard, and chipboard to desk, counter, and cabinet tops. Also used for laminating with sheet metals or wood veneers.

MANUFACTURER'S STATEMENT

Roberts Earthbond Adhesives release no hazardous vapours, which means that installations can be completed during regular business hours. The products are suited for use in health care facilities, and other commercial applications where hazardous vapors may present potential health risks. The products do not contribute to sick building syndrome. While providing environmentally safe products, quality does not suffer with these adhesives. They represent premium quality, high performance adhesives.

PRODUCTION

PACKING / SHIPPING

MINIMUM,
RECYCLED,
RECYCLABLE
PACKAGING

INSTALLATION / USE

MINIMUM
INSTALLATION
HAZARDS

LOW TOXIC
EMISSIONS
IN USE

RESOURCE RECOVERY

OTHER

RESEARCH &
EDUCATION
PROGRAMS

PRODUCT
Miscellaneous low toxicity adhesives

MANUFACTURER
W. F. Taylor Co., Inc.
13660 Excelsior Dr.
Santa Fe Springs, CA 90670
Tel (310) 802-1896, (800) 397-4583
Fax (310) 802-2831

DESCRIPTION
ENVIROTEC HEALTHGUARD ADHESIVES: Contain no solvents, no alcohol, no ammonia and no hazardous chemicals.

2027 PRESSURE SENSITIVE ADHESIVE, 2030 VERSA-BOND, 2045 BOND-N-PEEL SYSTEM FLOOR SPREAD, 2055 BOND-N-PEEL SYSTEM TOP BOND ADHESIVE, 2056 POWER-TAC: Releasable and rebondable pressure-sensitive adhesives, for direct glue-down of carpets and resilient floorings. Flooring materials may be lifted and reset or replaced. Adhesives remain permanently tacky, and will rebond upon contact. They withstand movement due to traffic and temperature changes.

2033 CLEAR THIN SPREAD TILE ADHESIVE, 2040 COVE BASE ADHESIVE, 2080 PREMIUM MULTI-PURPOSE ADHESIVE, 2090 VINYL BACKED ADHESIVE, 2093 RUBBER FLOORING ADHESIVE, 2087 EVR WHITE SHEET GOODS ADHESIVE: Fast-drying adhesives that will not contribute to staining or shrinkage. For use with wide variety of flooring materials.

2060 CONTRACT MULTI-PURPOSE ADHESIVE, 2070 MULTI-PURPOSE ADHESIVE, 2072 MULTI-PURPOSE LATEX ADHESIVE: Strong, wet bond, for hard-to-bond materials.

2095 TUB KIT AND CONSTRUCTION ADHESIVE: High strength adhesive to install high pressure laminate tub kits. Can also be used on plywood, paneling, ceramic tile, corkboard, insulation panels, and other rigid panels when applied to a rigid surface.

WB-100 WATERBASED CONTACT ADHESIVE: For bonding plastic laminate, wood, metal, fabric, leather, wallboard, linoleum, ceramics, glass, and rubber.

MANUFACTURER'S STATEMENT
All adhesives conform to current OSHA and EPA standards. The products contain no solvents, no alcohol, no glycol, no ammonia, no carcinogens, and release no VOCs after curing at room temperature. Antimicrobial protection of wet and dry films is a standard feature.

Extensive air quality testing has been performed on products. Envirotec Healthguard Adhesives were used to install 30,000 yards of carpet in the EPA headquarters, and tests performed during installation indicated significantly lower rate of emission than the EPA proposed emission levels.

Adhesives are packaged in strong, corrugated, recyclable cartons, which are made of recycled paper, containing two 2½-gallon pouches with reusable spout in 5-gallon pails.

PRODUCT
Low toxicity epoxy adhesives

MANUFACTURER

Laticrete International, Inc.
1 Laticrete Park N.
Bethany, CT 06525-3498
Tel (203) 393-0010, (800) 243-4788
Fax (203) 393-1684

Ontario:
Tel (519) 576-9840
Fax (519) 745-3822
Quebec:
Tel (514) 346-0107
Fax (514) 346-4428
Western Canada:
Tel (604) 732-1883
Fax (604) 732-5556

DESCRIPTION

LATAMASTIC 15: White, nonflammable latex adhesive designed for interior installations of ceramic tile on both floors and walls. Water resistant, for use over cement backerboard, gypsum wallboard and plywood. Low odor, water resistant and easy to use. For interior use only.

LATAPOXY 300: Complete, ready-to-mix epoxy installation system specifically formulated to install natural and agglomerated marble and granite tiles. Nonflammable and easily cleaned with water while fresh. Consists of 2 liquids and 1 powder component.

LATAPOXY 210: High strength epoxy adhesive designed for the installation of ceramic tile on sound, clean surfaces, such as exterior grade plywood, concrete, block, brick, steel, vinyl tile, and asphalt tile. Ideally suited for heavy duty commercial installations where materials must resist physical abuse, shock, and chemical attack. Easy to use, low toxicity, nonflammable and water cleanable. High bond strength, fast setting, durable and chemical resistant. For the installation of ceramics, brick, slate, tile, pavers, stone, marble, and quarry tile.

LOW TOXIC EMISSIONS IN USE

DURABLE

MANUFACTURER'S STATEMENT

See p. 114.

WALL COVERINGS

The use of wall coverings has a long history in interior decor, and has evolved from loosely hanging tapestries to laminated mylars. Though wall coverings are relatively durable, they are usually changed as fashion dictates, long before the end of their practical life.

MATERIALS

FIBERS

Both natural and synthetic fiber wall coverings are made in many different styles: "strings" of fibers can be laid parallel directly onto the paper backing; jute, sisal, cellulosics, or synthetics are woven; and a variety of ready-made fabrics, such as silk, cotton, and polyester are also used. The paper backing gives strength and allows more convenient installation.

As an alternate to paper backing, fabrics are sometimes applied as a paneled and padded wall, introducing upholstery foam into the combination, which adds to environmental and health impacts. (See FURNITURE, p. 216, for information on upholstery foams.)

Scrim, a loosely woven fabric, is used as a backing to add dimensional stability to commercial vinyl wall coverings. It is an essential component of this class of wall covering, and is made of cotton, polyester, or a blend. (See CARPETING, pp. 51, 52, and FURNITURE, pp. 212–14, for more information about the impact of manufacturing these materials.)

VINYL

Vinyl, used as the face material for both commercial and residential wall coverings, may have either a scrim backing, as mentioned above, or a paper backing. While manufacturers promote a heavier "film," or facing, for strength and resilience, stronger vinyls are now available that use less material and have better wear resistance. Meanwhile, industry and consumer standards are slower to change than product development, and in many applications these new lightweight, efficient products have not been given the chance to prove themselves.

All vinyls are made from petroleum, using a polymerization process, which creates serious risk of chemical exposure for workers, and produces hazardous waste. Cadmium, a hazardous heavy metal, is used as a heat stabilizer during the processing of vinyl films. The product remains soft with the use of plasticizers. It is chemically unstable, and may offgas for long periods. Chemical biocides are also a common ingredient. They may be added to the liquid vinyl, where they become a permanent component of the product, or applied as a topical treatment. Both methods result in additional risk of chemical exposure to workers and to users.

Color is often an integral part of the final product, and heavy metals, such as cadmium and lead, have traditionally been used in the pigments, though they are now severely restricted by law. Patterns can also be printed over the background. Solvent- or oil-based inks are generally used for this, although a few manufacturers are beginning to use safer vegetable-based or water-based inks. Emissions from the inks are a source of indoor air pollution for long periods.

PLASTER

Plaster is a gypsum product that is mixed with water. It is a traditional building material, but modern additives such as polyvinyl, acrylic, and mineral fillers (perlite, vermiculite, blast furnace slag, or wood fiber) have improved plaster's workability and other characteristics. Prior to being banned, asbestos was also used as a filler, and may present a danger from dust and flakes in older buildings.

Plaster-covered fabric wall coverings have also been developed, which give a wall surface texture, while hiding flaws in old substrates. One type combines plaster and jute, and has been included in the product sheets. (See INTERIOR CONSTRUCTION PANELS, p. 36, for more information about gypsum production.)

PAPER

Paper production, though from a natural, potentially renewable, raw material source, has major environmental impacts at practically every stage. Logging practices—forest monoculture, clear-cutting, and deforestation leading to land erosion and loss of habitat—have become one of the world's most pressing environmental issues. Pulp mill effluent is extremely hazardous to marine life, and stack emissions damage forests, are a health risk, and add to acid rain. Paper-making processes utilize hundreds of chemicals, with a resulting toxic waste water problem.

Recycled papers are now widely available for printing purposes and for building products. The use of recycled paper fiber greatly decreases the demand on forests, the energy needed, and the pollution produced in making paper. Continent-wide paper recycling programs are amassing vast quantities of post-consumer paper, as well as post-industrial waste. This available resource needs to be utilized, but there are not enough industries yet capable of receiving it and converting it into useful products. As mentioned above, inks present an indoor air problem, as well as having a negative environmental impact. Virtually all decorative wallpapers are printed or colored, although a few simply provide a textured surface that will accept paint.

Traditional embossed paper wall covering, known as Anaglypta, is one such product that requires painting. A similar product, Lincrusta, is made from putty, with a high linseed oil content and odor. It remains pliable and "fresh" until given a coat of paint, which initiates oxidization.

FOILS

Foil wall coverings contain only very small quantities of metal, and their impact is primarily the same as for paper. Although it would be possible to recycle the metal once stripped from the wall, since metal recycling can burn off any residual paper backing, it would be neither a practical nor economic proposition, considering the small quantities to be recovered. (See FURNITURE, p. 210, for information about metal production.)

MYLAR

The metal content in metallized mylar wall covering is even more minimal than that in foils. It occurs only as a molecular coating, and an ounce of metal goes a very long way. The largest components of this type of wall covering are plastic and paper. (See FURNITURE, p. 209, for more information about plastics.)

MANUFACTURING PRACTICES

With more stringent government-imposed restrictions on effluent and emissions from plants, many of the smaller, long-established manufacturers are financially unable to upgrade their facilities to the new standards. They are succumbing to corporate takeovers or massive plant shutdowns resulting in worker layoffs. Technological advances, computerization, and state-of-the-art facilities, while reducing environmental impact, have also meant that high production can be maintained with little labor. This has far-reaching social ramifications.

Additionally, as in the carpet industry, there is a growing concentration of wall covering manufacturers in a small geographical area, leading to overused resources.

INSTALLATION METHODS

Wall covering installation can be a relatively nonhazardous activity, requiring very few installation materials, of which low toxic types are readily available.

Sizing, a treatment that renders the bare wall less absorbent to adhesive, is a relatively safe material, consisting of water and starch, and small amounts of mildew retardants. (See INSTALLATION MATERIALS, pp. 106, 116–21 for further information on wallpaper pastes.)

Dry-glue systems, such as prepasted wallpapers, and those with a self-adhesive backing create less adhesive waste and cause less exposure than wet adhesive methods, as the installer is not involved in mixing or handling adhesives. The self-adhesive backing, however, creates waste paper or plastic.

Traditional water and starch paste is still readily available and widely used, being recommended for most paper and paper-backed wall coverings. Water-based wall covering glues are subject to mold and mildew growth, and insect attack. As a result, the addition of insecticides and biocides is a normal procedure, though low toxicity types are available. Solvent-based adhesives offgas toxic emissions during installation when the compounds are fresh, but also through porous textile or paper wall coverings for prolonged periods. (See INSTALLATION MATERIALS, pp. 105–7, for further information.)

MAINTENANCE

While some varieties of fibrous wall coverings will shed their own fibers, most uncoated, porous, textile wall coverings will, in addition, attract, accumulate, and then release dust particles, bacteria, and odors and gases found indoors. Only regular vacuuming can minimize the emissions. This condition is chronic during the life of the product.

With the advent of highly chemical-resistant synthetic fibres, such as polyolefin, manufacturers promote the benefits of cleaning with industrial-strength chemical cleaners or bleaching to kill bacteria and prevent odors and stains, thereby sanctioning the use of these toxic agents. If rigorous cleaning is required for a given installation, perhaps fiber wall covering is not the appropriate material to use.

The main characteristics for an easily maintainable wall covering are its washability, surface texture, and water resistance.

RESOURCE RECOVERY

Because of the nature of wall covering, it is not practically recyclable. The combination of materials (facings and backings) leaves a waste product that cannot be separated, even though the material may be strippable from the wall. Attempts have been made to develop a biodegradable formulation for vinyl wall covering, though this also reduces the desirable attributes of durability and strength. As with biodegradable plastic, although the concept is optimistic, the reality of North American landfills is that no light ever reaches the material to accomplish the molecular breakdown.

PRODUCTION

SUSTAINABLY
ACQUIRED, OR
RENEWABLE
RESOURCE

LOW
EMISSIONS
PLANT

PACKING / SHIPPING

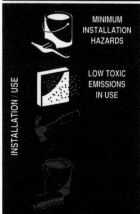

INSTALLATION / USE

MINIMUM
INSTALLATION
HAZARDS

LOW TOXIC
EMISSIONS
IN USE

RESOURCE RECOVERY

OTHER

RESEARCH &
EDUCATION
PROGRAMS

PRODUCT
All natural wall texturing material

MANUFACTURER
Auro Organic Paints (Germany)
Imported by Sinan Co., Natural Building Materials
P.O. Box 857
Davis, CA 95617-0857
Tel (916) 753-3104

RELATED PRODUCTS
Auro paint, p. 145
Auro adhesive, p. 116

DESCRIPTION
NO. 311 NATURAL FIBER WALL TEXTURE: Ready-to-use interior plaster from plant fiber materials and binding agents. Suitable for all absorbent surfaces, but not large areas of masonry.

MANUFACTURER'S STATEMENT
Auro products contain no isoparrafinic hydrocarbons (petroleum solvents), which are present in many other products.

Auro has done years of research and development, production and processing of natural substances to arrive at their products. In close cooperation with experienced craftspeople, quality is constantly controlled and improved where possible. Auro was the first producer worldwide to list all ingredients of their environmentally clean products. Plant chemistry's raw materials come from a closed ecological cycle which makes them permanently available.

Auro intentionally maintains small manufacturing plants, producing products for international distribution. This strategy supports local economies and maintains a high quality of product.

Sinan's catalog/newsletter is printed with soybean inks on 100% recycled paper, that has neither been de-inked or bleached.

PRODUCT
Natural fiber wall covering materials

SUSTAINABLY
ACQUIRED, OR
RENEWABLE
RESOURCE

PRODUCTION

MANUFACTURER
Merida Meridian Inc.
360 W. Genessee St.
Syracuse, NY 13201-1071
Tel (315) 422-4921

RELATED PRODUCTS
SISAL FIBER PROTECTION: Spray-on petroleum-based chemical containing polymers (and trichlorethylene) that seal fibers against water spotting, soils, and stains. Helps keep sisal's original colors lightfast. Also for use with coirs and wools.

Adhesives, pp. 105–6, 116–23

DESCRIPTION
TEXTURWORKS: 100% woven sisal wall covering line, ranging from finer to rougher structures. Natural and dyed or patterned sisal, with a solid natural latex or paper backing. Many colors and patterns available. Width: 48" or 39"; weight varies according to style of weave. Many are pretrimmed to facilitate installation.

HIMALAYA: 100% woven coir (57% coco, 43% sisal) wall covering, with a solid natural latex backing. Width: 39"; weight: 79 oz. per lin. yd. Pretrimmed.

MANUFACTURER'S STATEMENT
Merida Meridian imports only natural fiber products, which have minimal environmental impact. Yarns are spun and woven on Jacquard looms in Europe. (See also **FLOORING**, p. 67.)

NOTE
Sealers and adhesives are toxic products, particularly during installation. Alternative low toxicity adhesives are listed on pp. 116–23. Low toxicity soil and water repellents may be available, although some experimentation and consultation with manufacturers may be required.

PACKING / SHIPPING

SEE PAGE 126 — MINIMUM INSTALLATION HAZARDS

LOW TOXIC EMISSIONS IN USE

DURABLE

INSTALLATION / USE

RESOURCE RECOVERY

OTHER

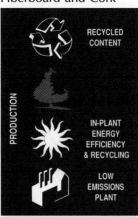

RECYCLED
CONTENT

PRODUCTION

IN-PLANT
ENERGY
EFFICIENCY
& RECYCLING

LOW
EMISSIONS
PLANT

PACKING / SHIPPING

MINIMUM
INSTALLATION
HAZARDS

SEE PAGE 126

LOW TOXIC
EMISSIONS
IN USE

INSTALLATION / USE

RESOURCE RECOVERY

OTHER

RESEARCH &
EDUCATION
PROGRAMS

PRODUCT
Decorative acoustical wall paneling

MANUFACTURER

Homasote Company
P.O. Box 7240
West Trenton, NJ 08628-0240
Tel (609) 883-3300
Fax (609) 530-1584

Eastern Canada:
(207) 427-6914
Western Canada:
(403) 256-4474
Ontario:
(416) 856-4994

RELATED PRODUCTS
Adhesives, pp. 105–6, 116–23

DESCRIPTION

DESIGNWALL™: Tackable, decorative, sound-absorbing paneling. Fabric (Guilford of Maine FR 701) laminated to ½″ structural Homasote fiberboard. Available in quartz, cherry, neutral, blue neutral and gray mix. Standard size: 4′ × 8′ × ½″.

NOVA CORK: Decorative, tackable cork panels where appearance and durability are prime considerations. A veneer of natural color, virgin cork laminated to Homasote structural fiberboard. Standard size: 4′ × 8′ × ½″.

440 SOUND-A-SOTE™: Use behind sheetrock or other paneling in walls or ceilings to reduce sound transmission through wall or floor/ceiling assemblies. Homasote structural fiberboard. Standard sizes: 4′ × 8′, 4′ × 10′, 4′ × 12′. Thickness: ½″.

440 HOMASOTE: Fabric-wrapped panel, protection board, hobby board, bed board, train board, tackboard, sidewall sheathing, or for cutout letters. Standard sizes: 4′ × 8′, 4′ × 10′, 4′ × 12′. Special order sizes: 8′ × 6′, 8′ × 8′, 8′ × 10′, 8′ × 12′, 5′ × 8′, 2′ × 4′, 3′ × 4′, 4′ × 4′. Thicknesses: ½″ and ⅝″.

MANUFACTURER'S STATEMENT

Homasote is made from 100% recycled newspaper, with no formaldehyde or asbestos additives. Heavy metal content is less than 100 parts per million. It is fully biodegradable, except for products laminated with other materials.

Our manufacturing process dates back to 1909. Homasote building products help in the conservation of more that 1,370,000 trees per year, and eliminate more than 160,000,000 lbs. of solid waste annually.

Homasote products are 100% environmentally safe, contributing no contaminants or pollutants to the atmosphere. They are energy saving, nailable, and easy to install. Water removed from products during manufacturing is recycled. Processing equipment is state of the art: efficient and technically advanced. Constant product innovation is sustained through ongoing research and development.

PRODUCT
Wood chip wallpaper

MANUFACTURER
F. Erfurt & Son (Germany)
Imported by
Eco-House (1988) Inc.
P.O. Box 220, Stn. A
Fredericton, NB E3B 4Y9
Tel (506) 366-3529
Fax (506) 366-3577

Livos Plant Chemistry/ USA
1365 Rufina Cir.
Santa Fe, NM 97214
Tel (505) 438-3448

RELATED PRODUCTS
LIVOS #410 URA STAINING PASTES, p. 147
LIVOS #540 LAVO WALLPAPER GLUE, p. 117
LIVOS #400 OR #405 DUBRO WHITE WALL PAINT, p. 147

DESCRIPTION
#1032 NATURE-FLEX WOOD CHIP WALLPAPER: Random texture created by adding fine wood chips to the paper. For use on walls or ceilings, has imperfection-hiding qualities.

Wallpaper is to be painted, and can be repainted up to 6 times. Roll size 21" wide × 410' long. Random pattern eliminates waste due to pattern matching.

MANUFACTURER'S STATEMENT
Nature-Flex has been awarded the German Ecologo, because it is made of natural fibres and de-inked, recycled paper. It has also been tested and approved by BBP, a federal German association which surveys and recommends healthful building materials for housing.

PRODUCTION

RECYCLED CONTENT

SUSTAINABLY ACQUIRED, OR RENEWABLE RESOURCE

PACKING / SHIPPING

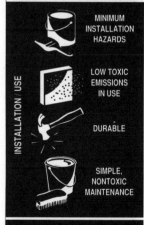

INSTALLATION / USE

MINIMUM INSTALLATION HAZARDS

LOW TOXIC EMISSIONS IN USE

DURABLE

SIMPLE, NONTOXIC MAINTENANCE

RESOURCE RECOVERY

OTHER

RESEARCH & EDUCATION PROGRAMS

SUSTAINABLY
ACQUIRED, OR
RENEWABLE
RESOURCE

PRODUCT
Gypsum-coated wall fabric

MANUFACTURER

Flexi-Wall Systems
(Division of Wall and
   Floor Treatments, Inc.)
207 Anderson Dr., P.O. Box 88
Liberty, SC 29657
Tel (803) 855-0500
Fax (803) 843-9318

Canadian Plaster-in-a-Roll
   Corp.
3721 Cadboro Bay Rd.
Victoria, BC V8P 5E3
Tel (604) 477-7631

RELATED PRODUCTS

FLEXI-WALL #500 ADHESIVE: Adhesive forms tough bond and hardens the Flexi-Wall coated fabric. Specially formulated acetate emulsion in a water base. Does not contain volatile solvents.

ANTI-G #400 SEALER: Water-based acrylic emulsion that will produce a clear, low gloss finish. Protects surfaces from most stains. Recommended for use on concrete block, marble, granite, finished or painted wood, metal, or vinyl wall coverings.

DESCRIPTION

PLASTER-IN-A-ROLL: High density, gypsum-coated, decorative, natural fabric that can be applied to any rigid surface. Supplied in 48" wide rolls, containing approx. 40 sq. yds. Length of roll is 30 yds.

Scotland Weave pattern available in 19 colors with a natural, stainless jute backing. Recommended for concrete block or other rough surfaces. *Indian Jute Weave* pattern available in 3 colors. Recommended for relatively smooth walls. *French Weave* pattern available in 25 special order colors. Recommended for smooth surfaces only. Custom colors are available for all products.

A factory-applied clear coating may be specified for added washability. In addition, an antigraffiti field-applied coating is available that provides a non-stick, clear, low gloss finish resistant to most stains. (Anti-G Coating #400). The surface can be coated with ordinary paint (for design flexibility) and will maintain the original textural effect. Class A fire rating, with no toxic smoke in the event of fire.

FASTER PLASTER: Heavy duty, flexible, white, plaster-finish wall liner used as a bridging material, underliner for wall coverings, or paintable veneer plaster coat.

MANUFACTURER'S STATEMENT

Plaster-in-a-Roll contains no asbestos or lead. It provides an excellent finish for masonry walls used in passive solar-heating systems. It is an effective barrier in covering existing lead paint. As Plaster-in-a-Roll breathes, it helps eliminate mildew problems and reduce delamination in humid areas.

NOTE

Adhesives and coating sealers may increase risk of exposure to installers. See INSTALLATION MATERIALS and FINISHES for possible low toxicity alternatives. Check with manufacturer for specific application.

SEE PAGE 126

MINIMUM
INSTALLATION
HAZARDS

LOW TOXIC
EMISSIONS
IN USE

DURABLE

SIMPLE,
NONTOXIC
MAINTENANCE

PRODUCT
Seamless, hard-surface wall finish system

MANUFACTURER

General Polymers
145 Caldwell Dr.
Cincinnati, OH 45216
Tel (513) 761-0011
Fax (513) 761-4496

General Polymers Canada
405 The West Mall, Suite 700
Etobicoke, ON M9C 5J1
Tel (416) 621-3988

DESCRIPTION

MACROSEPTIC®: Seamless, hard surface wall finish system, with Intersept® to inhibit the growth of harmful bacteria and fungi.

Suited for working environments that demand high performance and hygienic freshness: research labs, hospital operating rooms, food processing plants, grocery stores, kitchens, locker rooms, showers, restrooms, cafeterias, animal housing, and manufacturing.

Hi-Build Wall Systems: With fabric reinforced scrims, can be applied to poured concrete or concrete block walls for smooth, easily maintainable wall finish. Impact resistant, dimensionally stable, impervious, tile-like finish.

MANUFACTURER'S STATEMENT

General Polymers has been developing composition resin flooring and wall finishing systems to solve the challenges presented by specialized environments. Chemists and technicians use state-of-the-art polymer technology in research and development to meet the broadest array of needs.

General Polymers is a partner of the Envirosense™ consortium who provide products manufactured with Intersept®, a low toxicity, broad spectrum biocide.

NOTE

The addition of biocides to materials may not be an appropriate practice for many applications. Biocides are not a substitute for adequate cleaning and maintenance.

PRODUCTION

IN-PLANT ENERGY EFFICIENCY & RECYCLING

PACKING / SHIPPING

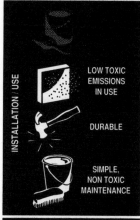

INSTALLATION / USE

LOW TOXIC EMISSIONS IN USE

DURABLE

SIMPLE, NON TOXIC MAINTENANCE

RESOURCE RECOVERY

OTHER

RESEARCH & EDUCATION PROGRAMS

PRODUCTION

PACKING / SHIPPING

INSTALLATION / USE

SEE PAGE 126

MINIMUM
INSTALLATION
HAZARDS

DURABLE

SIMPLE,
NONTOXIC
MAINTENANCE

RESOURCE RECOVERY

OTHER

PRODUCT
Fabric-backed vinyl wall covering

MANUFACTURER
Columbus Coated Fabrics, Borden, Inc.
1280 North Grant Avenue,
Columbus, OH 43201
Tel (614) 297-2906

RELATED PRODUCTS
Adhesives, pp. 105–6, 116–23

DESCRIPTION
GUARD CONTRACT WALLCOVERINGS: Full range of contract vinyl wall coverings. Reformulated as of January 1, 1991, to eliminate the use of cadmium in pigments and as vinyl stabilizers. A proprietary combination of barium and zinc has been used in place of cadmium. Uses a nonmercury-based biocide to inhibit mildew, fungal, and bacterial growth.

Guard-plus Series™ contains a stain-resistant protective coating, which allows cleaning with soapy water, pine oil cleaners, or 50% rubbing alcohol/water solution.

Class A flame retardant classification for all products. 5-year warranty given for all products.

MANUFACTURER'S STATEMENT
A new Guard logo, a global view of the world, has been selected to designate current and future products that reflect the commitment to the principles of environmental responsibility set forth by Borden, Inc.

NOTE
The addition of biocides to materials may not be an appropriate practice for many applications. Biocides are not a substitute for adequate cleaning and maintenance.

PRODUCT
Fabric-backed vinyl wall covering

MANUFACTURER
GenCorp Polymer Products
3 University Plaza, Suite 200
Hackensack, NJ 07601
Tel (201) 489-0100
Fax (201) 489-4394

RELATED PRODUCTS
Adhesives, pp. 105–6, 116–23

DESCRIPTION
GENON CONTRACT WALL COVERINGS: Full range of contract and residential vinyl wall coverings. Embossed vinyl face with polypropylene scrim backing.

MANUFACTURER'S STATEMENT
Wall coverings are tested to assess the effect of landfill disposal and ground water leaching of chemicals from discarded vinyl material. Research is underway to continue investigating approaches to improve the future potential for the biodegradability of vinyl wall coverings.

GenCorp Polymer Products has a commitment to product safety and will continue to not only adhere to governmental standards, but persevere in the technological development of vinyl wall coverings that reduce VOC emissions.

Genon pledges proceeds from the sale of Naturally Genon wall coverings to the Student Conservation Association (SCA), the oldest nonprofit educational organization that provides full-time volunteers to work in over 250 national parks, forests, wildlife refuges, and other public land sites. With Genon's support, 30 additional students are able to work.

GenCorp's design center now utilizes water-based inks for all product development strike-off samples (typically 70% of these were silk-screened using solvent-based inks).

GenCorp participates in the Environmental and Technical Committees of the Wallcovering Manufacturers Association, the Chemical Fabrics and Film Association and the Gravure Association of America.

PRODUCTION

PACKING / SHIPPING

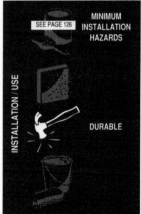

INSTALLATION / USE

SEE PAGE 126   MINIMUM INSTALLATION HAZARDS

DURABLE

RESOURCE RECOVERY

OTHER

RESEARCH & EDUCATION PROGRAMS

FINISHES

• PAINTS • STAINS • VARNISHES • SEALERS •

There are three categories of paints, stains, varnishes, and sealers that have been included in the Product Reports because of special qualities, such as low toxicity or low environmental impact. These are classed as "all natural," "low toxicity (water based)," and "low solvent, low biocide." Some other products have also been listed which have unique properties but do not fit readily into these categories. The broad range of other paints available which may meet some of the criteria is not represented here because information on their properties was not adequate for an assessment.

There are two major branches in paint technology, the solvent-based and the water-based types. Most of the products listed in this guide are water based. Solvent-based paints have been traditionally considered more durable, but they produce a great deal of hazardous vapor when curing and require cleanup with hazardous solvents. The volatile content contributes to local air pollution and the cleanup solvent poses a serious disposal problem. For these reasons, solvent-based consumer paints are being phased out slowly. Water-based paints are generally less hazardous to handle, but still contain a variety of toxic ingredients, as well as biocidal additives to prevent fungus growth and spoilage in the can. The important difference is that the materials used in water-based paints are much safer and less volatile than the petroleum spirits used in solvent-based paints, and they can be cleaned up with water. They contribute far less to local air pollution and hazardous waste problems.

Today there is almost no difference in performance between solvent-based and water-based paints for most indoor applications. Advances in chemistry have made durable finishes possible without most of the toxic solvents which were commonly used as little as ten years ago. Even high performance automotive finishes and transparent or opaque lacquers are now available in water-based varieties which contain far less volatile materials.

Other interesting developments in coating products are:
• Traditional paints made from milk protein (casein) have been reintroduced.
• Several water-based varnishes for hardwood floors are now available. They

are durable and much safer to handle than the traditional solvent-based urethanes and Swedish polymer oils.

- Very low toxicity, clear sealers are available which can be used as finishes for woodwork, or for sealing in volatile gases and odors from existing materials.

PAINTS

ALL NATURAL

The all-natural finishes are made by a small but diverse group of European manufacturers who share a common commitment to a philosophy of harmony with nature. The primary materials used in the finishes are extracts from plants, and minimally processed, earthen minerals such as chalk and iron oxide. It is important to emphasize that these products are not merely variations on modern paint chemistry, but are fundamentally different from the products available in the mainstream market. The approach is often called "plant chemistry." The recipes used in all-natural formulations are usually drawn from traditional methods and materials which have proven themselves over centuries of use.

The main environmental merit of all-natural finishes is that the raw materials are either from renewable plant sources or from abundant earth minerals. The steps used in processing these raw materials are also carefully evaluated to minimize toxic exposure, energy use, air and water pollution, and waste. The raw materials and processes have proven to be safe because they have been in use for many generations without negative health effects. This rationale is well recognized in toxicity evaluations and is called the Generally Regarded As Safe (GRAS) standard.

One distinction among manufacturers of all-natural finishes is that, while all make some oil-based paints, some use absolutely no petroleum-based materials, while others use small amounts of low odor petroleum solvents. Those who use no petroleum products usually use citrus oils, such as orange and lemon peel extracts, as solvents in formulations which require it. Because the odor of citrus oils is considered too strong by some users, some manufacturers substitute small amounts of a petroleum solvent which has been processed to remove the aromatic portions, and therefore has little odor. These solvents are called de-aromaticized isoparrafinics.

The majority of all-natural finishes do not require any special handling or application, and they perform very well in terms of coverage and hiding ability (the ability to cover another color). They are also as durable and washable as conventional finishes of the same class (such as gloss or semigloss). They do, however, generally cost two to three times more than their conventional counterparts, due both to the care and cost in making them, and to the fact that they are imported. Even at this higher price, paint is not a large cost item as a proportion of typical interior construction or renovation.

LOW TOXICITY

Some manufacturers have specialized in formulating low toxicity paints, both for those customers who are very aware of the environmental consequences and toxicity problems of paints, and for those who have special sensitivities and do not tolerate conventional paints well. Whereas the all-natural finishes are distinguished from conventional paints by what goes into them, the low toxicity finishes are distinguished by what does *not* go into them. These paints are variations of common formulations, such as latex emulsions and acrylics (See PAINT BASES, below.) The important difference is that such ingredients as solvents, stabilizers, dark pigments, and toxic metals have been either eliminated or replaced with safer materials. Virtually all of the low toxicity formulations are water dispersed and contain a minimal amount of volatile solvents. Some of the changes in formulation may make the paint 30% to 50% more expensive than conventional paints, and others may reduce the convenience of handling or, in some cases, reduce its performance slightly.

The low toxicity formulations are generally available in a range of types similar to conventional paints (matte, eggshell, and semigloss). Their application and performance are similar, though some may have less coverage and hiding ability.

One important distinction of low toxicity paints is that they are often sold in a limited range of colors, with further tinting left to the purchaser. The chemistry of tints is variable and complex, and the acceptability of odors from tints is a matter of individual sensitivity. If other tints are desired, the purchaser must use a tinting system available at paint stores, or purchase a nontoxic tint from the all-natural suppliers listed in the Product Reports, if there is concern about conventional tinting systems.

LOW SOLVENT, LOW BIOCIDE

Since the restriction of lead and cadmium in consumer paints (see TINTS, below), two of the major concerns remaining are the solvent content, and the biocide in water-based paints which retards spoilage and mildew growth. (See PRESERVATIVES, p. 140.) A few manufacturers will provide paints from their stock lines which have had all preservatives left out of the mixing process. This reduces their shelf life but makes the paints safer to handle and better tolerated by sensitive individuals. Biocide-free paints are available by special order for a small extra charge.

Lists of environmentally superior products, including paints, are available from such programs as the U.S. Green Cross, the Canadian Environmental Choice program, and the German Blue Angel program. Listed paints have met program standards for reduced volatile contents, and they contain none of the listed hazardous biocides.

SUMMARY OF PAINT PROPERTIES

1) "ALL NATURAL" EUROPEAN PAINTS

ADVANTAGES
- Carefully selected, low toxicity contents
- Made primarily from renewable resources (plant materials)
- Very durable and washable

DISADVANTAGES
- Two to three times the cost of conventional paints
- May have longer drying time

• Some limits on colors and available only from a few dealers or by mail order

2) DOMESTIC ''LOW TOXICITY'' PAINTS

ADVANTAGES

• May have even lower toxicity than low biocide paints (below)
• Modestly priced (30% to 50% more than conventional paints)

DISADVANTAGES

• Available only from a few dealers and by mail order
• Limited tints (tinting systems add to toxicity)

3) LOW SOLVENT/ LOW BIOCIDE PAINTS

ADVANTAGES

• Lower in volatile contents than conventional paints
• Do not contain the more hazardous biocides
• Available at paint stores as "listed products" for no extra cost

DISADVANTAGES

• Not as rigorously formulated as #1 and #2 above

COMPOSITION OF PAINTS

PAINT BASES

The paint vehicle of a solvent-based paint is a binder made from oils or resins dissolved in a solvent (usually petroleum spirits) with pigments added. The binders of water-based paints are resins such as latex, synthetic latex (styrene-butadiene), acrylics, or vinyls such as polyvinyl acetate. These binders are mixed in water with agents (called dispersants) to maintain them in suspension. A wide range of other agents is also added to water-based paints to make them apply more easily, cover better, dry faster, and to give them various finishes ranging from matte to gloss.

Whereas a solvent-based paint is normally glossy and produces a strong "film" on the painted surface, a water-based paint is naturally matte and tends to penetrate porous surfaces. In order to give water-based paints a glossy finish, such agents as coalescing solvents are used. These add to the chemical complexity of the paint and usually also add to toxicity.

Almost all of the materials used in conventional paint bases are derived from processed petroleum, and they cause air and water pollution and hazardous waste in their production. One of the few exceptions is the use of some plant oils, such as linseed and soybean oil, as "drying oils" for solvent-based, alkyd paints. These are oils which are treated so that they will harden when exposed to air. There is a large, untapped potential in the domestic paint industry for using seed oils in paints and coatings.

TINTS

The main tint in most interior paints is an opaque white, which provides the hiding ability. Colored tints are then added as desired. The white base in many paints is zinc oxide or titanium dioxide. Zinc oxide is produced from simple oxidation of zinc metal, and it is a low toxicity material. The main pollution associated with it is caused by the mining and smelting of zinc. In the past, it was difficult to be sure that zinc products were not contaminated with hazardous lead, but refining has now advanced to the point where lead-free zinc is widely available. Titanium dioxide is the most effective white pigment for paints, but its production requires sulfuric acid and chlorine treatments, which produce large amounts of toxic waste. Titanium manufacturers are now improving their acid process recycling and waste reduction to minimize this problem. Conscientious paint manufacturers buy from suppliers with the best waste controls.

Colored pigments are made from a wide variety of chemicals, ranging from relatively safe mineral oxides and plant pigments to very toxic lead and cadmium compounds. Lead and cadmium are both heavy metals, which are now closely regulated in consumer products due to health risks. Lead is clearly linked to neurological damage and will accumulate in persons who are chronically exposed. Children are especially at risk because they can suffer learning disabilities after exposure. Cadmium is clearly linked to kidney damage. Lead has been limited to very low amounts in interior paints since the 1970s, but the regulation of cadmium and other hazardous metals, such as chromium, in tints dates only from the 1980s.

A good general rule for low toxicity paints is that the tint is likely to be the component with the greatest environmental impact and health risks. For this reason pastel shades are preferable to more heavily saturated colors.

Health Note: Older buildings may contain lead paint, which creates a hazard when it begins to peel, or when paint removal is attempted during renovation. Check with local health authorities or your Environmental Protection Agency (EPA) regional office for advice. Also, exterior paints and coatings, and industrial paints may still contain toxic heavy metals. Do not use these for interior finishing.

PRESERVATIVES

Water-based paints have two types of spoilage problem: they will tend to turn sour or grow fungus in the container (usually after being opened), and they will support fungus (mildew) where used in damp locations. In order to control this problem, manufacturers add biocides during manufacture which are toxic to bacteria and fungi. Until the late 1980s, one of the most common biocides was a hazardous mercury compound called phenyl mercuric acetate (PMA). This biocide has now been banned for use in interior paints, and has been replaced by safer compounds containing no heavy metals. Several very toxic biocides including heavy metals are still used in exterior paints and coatings.

The majority of biocides can be avoided for interior paints. Water-based paint manufacturers who maintain careful control of their mixing are able to nearly eliminate the need for preservatives to prevent spoilage in the can. The only shortcoming is that shelf life may be reduced, particularly after opening, making it necessary to use up these paints within a few weeks after opening.

The need for fungus control of painted surfaces in damp locations can be reduced by proper insulation and ventilation, and by choosing a durable gloss or semigloss paint.

Exposure while painting is a serious health risk if paints contain toxic and volatile ingredients. Most exposure to the painter, and others in the building, occurs through the inhalation of vapors released as the paint cures. The painter is also exposed by skin contact. It is very difficult to protect the painter from toxic vapors because only industrial respirator masks containing carbon filters can absorb vapors. These are expensive, require maintenance, and are uncomfortable to wear. It is nearly impossible to protect other building occupants. Skin contact can largely be avoided by wearing surgical gloves, though many professional painters refuse to do so because they will not acknowledge the risks. These problems underscore the importance of using safer, low toxicity or all-natural paint formulations.

Within a few hours or days after painting, the emissions from paints decline, but all paints will continue to release trace amounts of gases for months after application. This is true for water-based paints as well as solvent-based paints. These long-term emissions also emphasize the importance of low toxicity or all-natural formulations.

Solvent-based paints containing petroleum spirits are a fire hazard. Those with very volatile solvents, such as lacquers, can be explosive during use. The method of application is also very important. Spraying releases large amounts of paint into the air, adding to inhalation and skin contact risk, as well as fire risk. Brush and roller application are far safer.

PACKAGING

Some paint manufacturers are choosing to label and box paint cans using recycled paper and cardboard products with simply printed labels. This should become the standard approach to packaging in the building products industries because, unlike consumer goods, packaging is a small part of product promotion. Simpler packaging costs less, and consumes less energy and resources, as well as producing less waste.

A few unique paint products, such as lime and milk paints, are shipped in dry powder form, and water is added to activate them. This reduces shipping costs and assures that the paint is not spoiled by extended time on the shelf (because it contains no biocides).

CLEANUP AND DISPOSAL

As discussed above, solvent-based paints should not be necessary for most indoor uses. If they are used, they should be handled by professional painters with brush and roller equipment, and the building should be vacated and aired for at least one week afterward.

Odor-reduced solvent is available for cleanup. If it is left to stand after use, the solids will settle to the bottom, and the clear solvent can be reused. The solids must be handled as hazardous waste. Most cities have hazardous waste receiving areas at their recycling facilities, which is where waste paint of any kind should go. If no receiving service is available, small residual quantities of paint can be left to evaporate and sent to waste pickup as a dry solid. This is, of course, only appropriate in areas where other disposal methods are unavailable. You should check with your local waste treatment center for information.

The best disposal method for paint is to have the right amount for the job so it is all used up. Remainders can be given to someone else to use, or stored for short periods in frost-free, well-ventilated conditions away from heat and flame.

MAINTENANCE

The washability and wear resistance of a paint are determined by the degree of hardness and gloss of the finish. These properties have been indicated in the Product Reports with the "durable" symbol for paints which are available in a semigloss or gloss finish. Matte finish, water-based paints are not very durable, and most cannot be washed with much success. Eggshell and semi-gloss, water-based paints are moderately durable and washable. The all-natural, oil-based paints are the most durable of those listed in this book, and are comparable to conventional alkyd paints.

STAINS

Stains are similar to paints in that there are both water-based and solvent-based varieties available. The primary difference between paints and stains is that stains contain fewer opaque fillers and resins, which form surface films. The tint in stains forms a much larger part of the product than for lightly colored paints. This makes the formulation of the tint critical.

Water-based stains are suitable for almost all indoor applications, particularly where they will be covered by a transparent coating of sealer or varnish, or protected from contact. Generally, the handling, application, cleanup, and disposal of stains is the same as for paints. Some specialty suppliers carry wood stains which contain no toxic ingredients and are colored entirely with natural minerals. These are the safest to handle.

VARNISHES AND SEALERS

There are many types of varnishes and sealers, among which are shellacs, resin varnishes, urethane varnishes, nitrocellulose lacquers, primer oils, acrylic sealers, penetrating/impregnating sealers, and moisture sealers. (Some of the industrial varieties are listed in the FURNITURE section under PAINTS AND FINISHES, p. 216–18.) This group of finishes is a very large one, and information is widely available on most types. For this discussion only a few options, used mainly for interior wood finishing, will be covered.

One of the most popular finishes for wood floors and interior woodwork is a solvent-based, urethane varnish. These are isocyanurates in a petroleum spririt solvent. Although it is toxic to handle (similar to solvent-based paints) this varnish is generally stable once cured. It does, of course, have all of the application, cleanup, and hazardous waste problems of solvent-based paints.

A safe and practical alternative is a water emulsion urethane, or acrylic-urethane, available as a one-part system which performs very well, is safe to handle and cleans up with water. These are also known as "water-based, penetrating polymer finishes," and are available in clear, or transparent colors. For heavier traffic areas and commercial use, a two-part system is available for professional application. The two-part system uses a catalyst which forms a "cross-linked polymer," making a very stable and durable finish. The handling, cleanup, and disposal of these water-based materials are much safer than for solvent-based products.

One of the most popular penetrating hardwood floor and cabinet finishes is a Swedish-type, formaldehyde-based polymer finish which produces large amounts of hazardous formaldehyde gas and other gases during application. It continues to release smaller amounts for several weeks after application. The popularity of this finish is due to the way it penetrates, hardens, and seals wood without producing a glossy finish, as do the varnishes and water-based urethanes described above. However, the installation health risk, and long-term air pollution potential of this finish make it undesirable for many.

The only finish with a similar appearance but without the air pollution problems is a Swedish or Danish penetrating, "drying oil" finish. These are usually based on linseed oil and waxes, and are available in "all natural" formulations. They are safe to handle, though solvent may be needed for cleaning tools. The drying oil finishes are less durable and require more maintenance than formaldehyde-based finishes. (See pp. 184–86.)

LOW TOXICITY SEALERS

Another important category of interior finishes for wood, porous tile, grout and mortar, gypsum board, and other porous surfaces are transparent, low toxicity sealers. These are similar to the water-based acrylic-urethane finishes discussed above, but they are formulated to penetrate and prevent water absorption, not to produce a durable surface finish. These sealers are also good for sealing materials, such as wood composition board, which emit odors and irritants, such as formaldehyde. They are well tolerated by people with sensitivities, and can be effective in reducing the characteristic odor of very old wood, which can be a problem in buildings with wood paneling.

FINISHES
All-natural Paint

PRODUCTION
SUSTAINABLY ACQUIRED, OR RENEWABLE RESOURCE

PACKING / SHIPPING
MINIMUM TRANSPORT ENERGY

INSTALLATION / USE
MINIMUM INSTALLATION HAZARDS

LOW TOXIC EMISSIONS IN USE

RESOURCE RECOVERY

OTHER

PRODUCT
Milk paint

MANUFACTURER

The Old Fashioned Milk Paint Co.
Box 222
Groton, MA 01450
Tel (508) 448-6336

Canadian Old Fashioned
  Milk Paint Co.
163 Queen St. E.
Toronto, ON M5A 1S1
Tel (416) 364-1393
Fax (416) 364-1170

DESCRIPTION

OLD-FASHIONED MILK PAINT: Dry, powder form paint using milk products, and mineral fillers and pigments (natural earth colors).

Produces a flat, grainy appearance—an authentic historic look. Adhesion to bare wood is practically permanent and paint strippers do not affect the finish. For use on wood and plaster.

Ingredients: lime, milk protein (casein), clay, and earth pigments such as ocher, umber, iron oxide, lampblack, etc. When wet, the paint has a slight milk odor which will disappear in several hours. The hydrated lime is strongly alkaline, but becomes totally inert after it dries. The other ingredients are inert inorganic materials.

Colors: barn red, pumpkin, mustard, bayberry, Lexington green, soldier blue, oyster white, pitch black.

MANUFACTURER'S STATEMENT

The product was redeveloped as the need for an authentic finish arose for handmade copies of early furniture from the pilgrim century to the time of the American Revolution. The product was commercialized after much research and investigation of old recipes and formulas, and trial and error.

No lead, chemical preservatives, fungicides, hydrocarbons, or other petroleum derivatives are used. Packaging in powder form reduces shipping costs, while ensuring that the product does not spoil in transit or on warehouse shelves.

Explicit directions for use are provided with the paint product, for best results.

NOTE

Milk paint should not be used in damp locations, as milk protein will support bacterial and fungal growth.

MANUFACTURER

Auro Organic Paints (Germany)
Imported by Sinan Co., Natural Building Materials
P.O. Box 857
Davis, CA 95617-0857
Tel (916) 753-3104

RELATED PRODUCTS

NO. 330 NATURAL COLOR CONCENTRATE: Fine dispersion of earthen and mineral pigments in natural binders. Eight rich colors for toning of white wall paints (No. 320 and 321). Can also be used diluted for glazing. Available in yellow ocher, Persian red, cobalt blue, umber brown, terracotta red, green, earthen brown, earthen black.

NO. 360 PLANT PIGMENTS FOR WALL GLAZING: Highly concentrated, pure plant pigments, dispersed in natural binders, for creative and colorful, bright and lively decoration of indoor walls. These are holistic, not pure, colors, also used as therapy. Available in yellow, red, blue, orange, red-violet, red (yellow tone), leaf green, brown, black, red (blue tone).

DESCRIPTION

NO. 320 ROOM WHITE PAINT: All-natural indoor wall paint. Contains no recycled titanium dioxide and has very low odor.

NO. 321 EMULSION WALL PAINT, WHITE: Indoor use, semigloss, very economical, good covering qualities, pleasant fragrance, easy application by roller. Breathing properties create pleasant room climate.

NO. 325 NATURAL CHALK PAINT: Inexpensive indoor wall paint powder, ready to mix with water, especially appropriate for basements, etc. Good choice for people with allergies. Off-white, spongeable, but not washable.

NO. 351 CASEIN WALL PAINT: White wall paint for interior applications. Not suitable for rooms with high humidity (kitchen and bathroom).

MANUFACTURER'S STATEMENT

Auro products contain no isoparrafinic hydrocarbons (petroleum solvents), which are present in many other products.

Auro has done years of research and development, production and processing of natural substances to arrive at their products. In close cooperation with experienced craftspeople, quality is constantly controlled and improved where possible. Auro was the first producer worldwide to list all ingredients of their environmentally clean products. Plant chemistry's raw materials come from a closed ecological cycle which makes them permanently available.

Auro intentionally maintains small manufacturing plants, producing products for international distribution. This strategy supports local economies and maintains a high quality of product.

Sinan's catalog/newsletter is printed with soybean inks on 100% recycled paper, that has neither been de-inked or bleached.

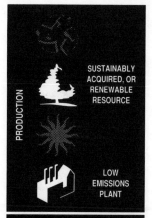

PRODUCTION

SUSTAINABLY ACQUIRED, OR RENEWABLE RESOURCE

LOW EMISSIONS PLANT

PACKING / SHIPPING

MINIMUM, RECYCLED, RECYCLABLE PACKAGING

INSTALLATION / USE

MINIMUM INSTALLATION HAZARDS

LOW TOXIC EMISSIONS IN USE

RESOURCE RECOVERY

OTHER

RESEARCH & EDUCATION PROGRAMS

PRODUCTION

SUSTAINABLY
ACQUIRED, OR
RENEWABLE
RESOURCE

LOW
EMISSIONS
PLANT

PACKING / SHIPPING

INSTALLATION / USE

MINIMUM
INSTALLATION
HAZARDS

LOW TOXIC
EMISSIONS
IN USE

DURABLE

RESOURCE RECOVERY

OTHER

PRODUCT
All-natural water-based wall paint

MANUFACTURER
Biofa Naturfarben GmbH (Germany)
Imported by Bau, Inc.
P.O. Box 190
Alton, NH 03839
Tel (603) 364-2400

RELATED PRODUCTS
BIOFA COLOR TINTS FOR WHITE WALL PAINT: Derived from clay and mineral pigments which do not contain lead, cadmium, or chromate. Colors available: red, orange, ochre, yellow, green, blue, brown, umbra, black.

DESCRIPTION
BIOFA WHITE WALL PAINT: Water-based interior paint made of natural resins, plant and ethereal oils. For use on plaster, concrete, rough fiber wallpaper, or over other conventional paints. Thick covering, matte finish, washable, quick drying times.

MANUFACTURER'S STATEMENT
Biofa was founded in 1980 and is dedicated to alleviating the present poisoning of the environment and its consequences to human life.

Biofa products use only raw materials which can be produced from nature and continually replenished, such as oils, resins, waxes, balsams, and plant pigments; which, through all-natural processes, acquire high quality technical properties; which after production can be returned to the natural cycle; which cause only minimal effects on workers and environments through production; and which do not threaten the user's health.

Bau had its beginnings as a residential design and construction company using nontoxic materials and finishes.

PRODUCT
All-natural water-based wall paints

MANUFACTURER
Livos Plant Chemistry (Germany)
Imported by

Eco-House (1988) Inc.
P.O. Box 220, Stn. A
Fredericton, NB E3B 4Y9
Tel (506) 366-3529
Fax (506) 366-3577

Livos Plant Chemistry / USA
1365 Rufina Cir.
Santa Fe, NM 97214
Tel (505) 438-3448

RELATED PRODUCTS
#410 URA STAINING PASTES: Staining pastes used for tinting water-based paints. Wide spectrum of pastel tones can be achieved (11 base colors).

DESCRIPTION
#400 DUBRO NATURAL RESIN WALL PAINT: Flat interior paint for use on stone and concrete walls, drywall, wallpaper and other porous, absorbent surfaces. Comes white and can be tinted with #410 Ura staining paste (water based).

#405 DUBRO NATURAL OIL WALL PAINT: Technically similar to #400, but contains less fragrant ingredients, for those who are also sensitive to certain natural ingredients, such as resins and essential plant oils.

All water-based products are normally shipped during nonfrost season only, to avoid shipping charges for heated service shipment. Products shipped November through April on request only and with an additional charge.

MANUFACTURER'S STATEMENT
Livos products are designed to improve the quality of your personal environment. All ingredients used are selected to meet the following criteria: 1) no health threatening materials; 2) ecological soundness of raw material production, manufacturing process, application, and disposal; 3) highest technical quality.

Livos products contain only natural and nontoxic ingredients, providing a pleasant fragrance during application and eliminating continued evaporation of fumes from the dried product. Finishes are harmless to humans, plants, soil, and water. Ingredients are known, natural and renewable, and require no testing on animals.

Livos maintains small manufacturing plants, producing products for local and international distribution. This strategy supports local economies and maintains a high quality of product.

Process acids are recycled and reused. Containers are of recyclable metal or cardboard/paper from recycled materials.

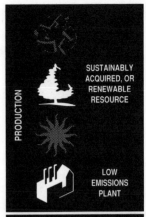

PRODUCTION

SUSTAINABLY ACQUIRED, OR RENEWABLE RESOURCE

LOW EMISSIONS PLANT

PACKING / SHIPPING

MINIMUM, RECYCLED, RECYCLABLE PACKAGING

MINIMUM TRANSPORT ENERGY

INSTALLATION / USE

MINIMUM INSTALLATION HAZARDS

LOW TOXIC EMISSIONS IN USE

DURABLE

RESOURCE RECOVERY

OTHER

RESEARCH & EDUCATION PROGRAMS

PRODUCTION

SUSTAINABLY
ACQUIRED, OR
RENEWABLE
RESOURCE

LOW
EMISSIONS
PLANT

PACKING / SHIPPING

MINIMUM,
RECYCLED,
RECYCLABLE
PACKAGING

INSTALLATION / USE

MINIMUM
INSTALLATION
HAZARDS

LOW TOXIC
EMISSIONS
IN USE

RESOURCE RECOVERY

OTHER

RESEARCH &
EDUCATION
PROGRAMS

PRODUCT
Miscellaneous natural, oil-based primers and top coats

MANUFACTURER
Auro Organic Paints (Germany)
Imported by Sinan Co., Natural Building Materials
P.O. Box 857
Davis, CA 95617-0857
Tel (916) 753-3104

DESCRIPTION
NO. 233 NATURAL WHITE UNDERCOAT: Suitable for priming or intermediate coats. For interior and exterior use on wood and neutral plaster surfaces.

NO. 234 NATURAL METAL PRIMER: Suitable for priming uncoated, rust-free steel and metal materials indoors and outdoors.

NO. 235 NATURAL RESIN OIL TOP COAT, WHITE: Semigloss lacquer with good covering qualities for indoors and outdoors. Apply by brush or spray gun. Slow drying to obtain permanently elastic surface.

NO. 236 NATURAL RESIN OIL TOP COAT, WHITE: Low-gloss lacquer for interior use only. Ideal for humid rooms. Slow drying to obtain permanently elastic surface.

NO. 237 NATURAL RESIN OIL RADIATOR PAINT: Special white high temperature lacquer for radiators. Creates pleasant room climate.

NO. 240 NATURAL RESIN-OIL TOP COAT, COLORED: Indoor/outdoor lacquer in 8 color tones: yellow ocher, Persian red, cobalt blue, umber brown, terracotta red, green, earthen brown, earthen black.

MANUFACTURER'S STATEMENT
See p. 145.

PRODUCT
Natural enamel wall paint and primer

MANUFACTURER
Biofa Naturfarben GmbH (Germany)
Imported by Bau, Inc.
P.O. Box 190
Alton, NH 03839
Tel (603) 364-2400

DESCRIPTION
Biofa wall paints are made of natural binders (tree resins, vegetable oils, and shellac). Pigments and fillers are from colored earth pigments and nontoxic metal oxides. The only preservatives used are pure ethereal oils and natural boric salts, making it possible to store unopened containers for approximately 1 year.

BIOFA WHITE PRIMING PAINT: Water repelling, breathable, dense white primer made from linseed and other plant oils, natural resins and fillers, and white pigments. For interior and exterior use on wood, steel, aluminum, and copper.

BIOFA NATURAL ENAMEL PAINT: Glossy, thick covering, dirt and water-repelling paint. Made from natural resins, linseed and other plant oils, and pigments. For interior and exterior use on wood, steel, aluminum, copper, etc. Especially suitable for protecting doors, garden furniture, windows, and railings.

MANUFACTURER'S STATEMENT
See p. 146.

PRODUCTION

SUSTAINABLY ACQUIRED, OR RENEWABLE RESOURCE

LOW EMISSIONS PLANT

PACKING / SHIPPING

INSTALLATION / USE

MINIMUM INSTALLATION HAZARDS

LOW TOXIC EMISSIONS IN USE

DURABLE

RESOURCE RECOVERY

OTHER

SUSTAINABLY
ACQUIRED, OR
RENEWABLE
RESOURCE

PRODUCTION

LOW
EMISSIONS
PLANT

PACKING / SHIPPING

MINIMUM,
RECYCLED,
RECYCLABLE
PACKAGING

INSTALLATION / USE

MINIMUM
INSTALLATION
HAZARDS

LOW TOXIC
EMISSIONS
IN USE

DURABLE

RESOURCE RECOVERY

OTHER

RESEARCH &
EDUCATION
PROGRAMS

PRODUCT
All-natural opaque enamel and oil paints

MANUFACTURER
Livos Plant Chemistry (Germany)
Imported by
Eco-House (1988) Inc.
P.O. Box 220, Stn. A
Fredericton, NB E3B 4Y9
Tel (506) 366-3529
Fax (506) 366-3577

Livos Plant Chemistry / USA
1365 Rufina Cir.
Santa Fe, NM 97214
Tel (505) 438-3448

RELATED PRODUCTS
#620 SKEIMA FLATTENING AGENT (OIL-BASED): Used to change finishes to semi-gloss, silky, or flat surface.

DESCRIPTION
#629 VINDO ENAMEL PAINT: Oil-based enamel paint for interior and exterior use on wood or metal. Excellent durability. Weather-resistant and gives tough, elastic, high gloss surface coat. Available in white, and 11 premixed, full-tone colors. Other colors can be achieved with #275 Leuko staining paste (p. 162).

#623 DURO METAL PRIMER: For interior and exterior applications. Provides elastic coating with excellent adhesive capacities. Low solvent content.

#665 AMELLOS SOLVENT-FREE OIL PAINT: Weather-resistant, elastic oil paint containing no solvents. For use on exterior surfaces which do not get much contact: fences, wood trims and fascias, animal habitats. Not for use on playground equipment, lawn furniture, windows, etc.

#625 CANTO WHITE ENAMEL (SATIN FINISH): Elastic natural resin enamel for interior use on wood and metal surfaces. Excellent for trim, doors and furniture.

MANUFACTURER'S STATEMENT
See p. 147.

PRODUCT
Miscellaneous all-natural oil paints

MANUFACTURER

Naturhaus (Germany)
Imported by
Lindquist Marketing
P.O. Box 3542
33118 Whidden Ave.
Mission, BC V2V 2T2
Tel (604) 826-1547

Lindquist Marketing
115 Garfield St.
Sumas, WA 98295
Tel (604) 826-1547
Fax (206) 988-2207

RELATED PRODUCTS

NATURAL THINNER #1, p. 194

DESCRIPTION

#03350/60 FENCE PAINT: Protection for all exterior wood surfaces. Deep penetrating qualities with good coverability. Elastic, resistant to steam, water, and ultraviolet light. Easy to clean without affecting the surface. Contains linseed oil, wood oil, natural resin, soya lecithin, orange turpentine, bentonite, and oxide pigments. Available in brown and black.

#05500 NATURAL METAL PRIMER/RUST PREVENTATIVE: Interior and exterior rust protection for iron. Lead-free, vapor-neutral paint with good coverability. Elastic and water resistant. Contains linseed oil, natural oils and resins, orange turpentine, bentonite, ferric oxide, mica, zinc oxide, soya lecithin, and natural driers. Available in matte brown.

#05510 EXTERIOR PRIMER: Primer for wood, metal, and concrete, for interior and exterior use. Temperature and ultraviolet light resistant. Contains linseed oil, natural oils and resins, titanium dioxide, mica, dolomite, talcum, orange turpentine and natural driers. Available in matte white.

#06000-60 INTERIOR AND EXTERIOR OIL PAINT: Fine glossy finish, elastic, thixotropic, and weatherproof oil paint. Contains linseed oil, natural oils, soya lecithin, bentonite, oxide pigments, soot, orange turpentine, titanium dioxide, and natural driers. Available in white, ocher, red, brown, black, green, and blue. Use with #05500 or #05510 Primers.

#06500 INTERIOR OIL-BASE PAINT: Fine glossy finish, elastic, color-stable oil paint for metal and cast iron. Contains linseed oil, natural oils, soya lecithin, bentonite, titanium dioxide, talcum, calcium carbonate, zinc oxide, orange turpentine, and natural driers. Available in white only.

#08000 INTERIOR PAINT: Good coverability, color stable, without titanium dioxide. The product is humidity resistant and easy to clean. Antistatic and free of preservatives. Contains chalk, lithopone, barite (heavy spar), lime, linseed oil, orange turpentine, pine oil, and natural driers. Available in white only. For use on concrete, drywall, gypsum, wallpaper, mortar, and plaster.

MANUFACTURER'S STATEMENT

See INSTALLATION MATERIALS, p. 118.

PRODUCTION

SUSTAINABLY ACQUIRED, OR RENEWABLE RESOURCE

LOW EMISSIONS PLANT

PACKING / SHIPPING

INSTALLATION / USE

MINIMUM INSTALLATION HAZARDS

LOW TOXIC EMISSIONS IN USE

DURABLE

RESOURCE RECOVERY

OTHER

FAIR BUSINESS PRACTICES

PRODUCTION

LOW
EMISSIONS
PLANT

PACKING / SHIPPING

MINIMUM,
RECYCLED,
RECYCLABLE
PACKAGING

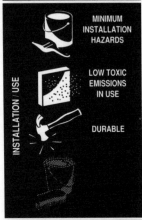

INSTALLATION / USE

MINIMUM
INSTALLATION
HAZARDS

LOW TOXIC
EMISSIONS
IN USE

DURABLE

RESOURCE RECOVERY

OTHER

RESEARCH &
EDUCATION
PROGRAMS

PRODUCT
Low toxicity water-based paints

MANUFACTURER
A.F.M. Enterprises Inc.
(American Formulating and Manufacturing)
1140 Stacy Court
Riverside, CA 92507
Tel (714) 781-6860, 781-6861   Fax (714) 781-6892

RELATED PRODUCTS
STRIPPER 66, p. 196
SUPER CLEAN, p. 200

DESCRIPTION
SAFECOAT PRIMER UNDERCOATER: Primes, protects, and seals imperfect substrates for walls and woodwork.

SAFECOAT ALL PURPOSE ENAMEL: Water-based gloss for exterior or interior. Finish coat for metal or wood. Excellent durability and scrub resistance.

SAFECOAT ENAMEL: Water-based semigloss allows easy brushing, fast dry, excellent gloss retention, hiding power, hardness, and scrub resistance.

SAFECOAT PAINT: Sealant and finish coat for walls and woodwork that cures to a flat and extremely water-resistant finish.

Products come in white or bone, and can be tinted to any light or medium shade, with any universal water-based tinting system.

MANUFACTURER'S STATEMENT
Toxicity tested by the Environmental Protection Agency and independent laboratories without the use of animals. Products found widely accepted by those with allergies and chemical sensitivity.

All products are formulated as nonhazardous and stable, and do not contain bactericides or fungicides that are classified as phenol mercury acetates, phenol phenates, or phenol formaldehyde. Contain no aromatic and aliphatic solvents which are present in many other products.

Cans and boxes are made from recycled materials.

PRODUCT
Low toxicity latex paint

MANUFACTURER

Benjamin Moore & Company
51 Chestnut Ridge Rd.
Montvale, NJ 07645
Tel (201) 573-9600
Fax (201) 573-1984

Benjamin Moore & Company Ltd.
139 Mulock Ave.
Toronto, ON M6N 1G9
Tel (416) 766-1173
Fax (416) 766-9677

DESCRIPTION

REGAL SERIES: Water-based emulsion, multipurpose interior latex paint. *Regal Wall Satin:* latex flat paint; *Regal AquaVelvet:* latex eggshell flat enamel; *Regal AquaPearl:* latex pearl finish enamel; *Regal AquaGlo:* latex satin finish enamel.

EXTERIOR LATEX: Water-based emulsion, multipurpose exterior latex paint. *MoorGard:* latex low luster house paint; *MoorGlo:* latex soft gloss house and trim paint;

MOORE'S LATEX ENAMEL UNDERBODY: Water-based emulsion, deep color base primer.

MANUFACTURER'S STATEMENT

Benjamin Moore Paints have met and surpassed Canadian government guidelines for many years, and now carry Environment Canada's Environmental Choice program EcoLogo for low pollution, low VOC emission paints.

PRODUCTION

PACKING / SHIPPING

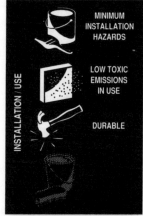

MINIMUM INSTALLATION HAZARDS

LOW TOXIC EMISSIONS IN USE

DURABLE

INSTALLATION / USE

RESOURCE RECOVERY

OTHER

PRODUCTION

PACKING / SHIPPING

INSTALLATION / USE

MINIMUM
INSTALLATION
HAZARDS

LOW TOXIC
EMISSIONS
IN USE

DURABLE

RESOURCE RECOVERY

OTHER

PRODUCT
Low toxicity latex paints

MANUFACTURER
Color Your World
10 Carson St.
Toronto, ON M8W 3R5
Tel (416) 259-3251   Fax (416) 259-4167

DESCRIPTION
WATER-BASED PAINTS: Water-based emulsion, multipurpose interior and exterior latex paints.

Interior Latex: flat, eggshell, satin, and semigloss; *Painter's Choice Latex:* flat and ceiling; *Designer's Touch Latex:* ceiling, flat, satin, and semigloss; *Velvet Pastels:* flat, eggshell, satin, and semigloss; *Interior Block Filler* and *Sealer Latex, Wallcovering All-prep, Interior Latex Sealer, Exterior Latex:* flat, satin, semigloss, and gloss; *Exterior Flat Acrylic Latex, Interior/Exterior Latex:* colors and semigloss; *Latex Floor, The Outsider-Metal Siding, Acrylic Driveway Sealer.*

MANUFACTURER'S STATEMENT
All products bear the Ecologo for water-based paints. Solvent-based stains which would qualify for the Reduced Solvent-Based Paint Ecologo are not yet available on the market. However, a line of high quality acrylic, solid color, water-based stains do bear the Ecologo and are a suitable alternative for most solid color stain applications.

PRODUCT
Water-based acrylic enamel spray paint

MANUFACTURER

Environmental Technology, Inc.
S. Bay Depot Rd.
Fields Landing, CA 95537
Tel (707) 443-9323
Fax (707) 443-7962

Environmental Technology, Inc.
 Canada
Unit 305, 2056 #10 Highway
Langley, BC V3A 6K8
Tel (604) 534-3810
Fax (604) 534-4173

RELATED PRODUCTS

#40509 RED OXIDE PRIMER

DESCRIPTION

ENVIROSPRAY ENAMEL SPRAY PAINT: All-purpose, durable enamel finish will adhere to most surfaces (plastic, metal, glass, ceramic, Styrofoam, fabric). For interior and exterior use. Available in wide range of colors.

Excellent one-coat coverage on nonporous surfaces, reduced overspray, soap and water cleanup, salt and weather resistant, will not crack or peel, high fading resistance.

Formulations contain nonchlorinated solvents, that are virtually nontoxic. Dimethyl ether is used as a propellant (ozone safe); no CFCs or freon are used. Nonflammable.

Long shelf life due to water-based system.

MANUFACTURER'S STATEMENT

ENVIROSpray is the only spray paint to be classified with the Blue Angel Symbol of the European Common Market, designated as an environmentally friendly product.

Contains no methylene chloride solvent, lead, ketones, hydrocarbons, aromatics, or other petroleum distillates.

PRODUCTION

PACKING / SHIPPING

INSTALLATION / USE

MINIMUM INSTALLATION HAZARDS

LOW TOXIC EMISSIONS IN USE

DURABLE

RESOURCE RECOVERY

OTHER

PRODUCTION

LOW
EMISSIONS
PLANT

PACKING / SHIPPING

MINIMUM
INSTALLATION
HAZARDS

LOW TOXIC
EMISSIONS
IN USE

DURABLE

SIMPLE,
NONTOXIC
MAINTENANCE

INSTALLATION / USE

RESOURCE RECOVERY

OTHER

PRODUCT
Interior and exterior paint

MANUFACTURER
Flecto Coatings
1000 45th St.
Oakland, CA 94608
Tel (510) 655-2470
Fax (510) 652-0969

Flecto Coatings
4260 Vanguard
Richmond, BC V6X 2P5
Tel (604) 273-8031
Fax (604) 273-6026

DESCRIPTION
DIAMOND "COLORS": Interior and exterior paint, in an interpenetrating polymer network base (acrylic). 10 colors available in gloss finishes (white and black available in gloss and satin finishes).

MANUFACTURER'S STATEMENT
Flecto's ongoing environmental goal is to achieve zero discharge. To date, they have reduced hazardous wastes by 80%, and the target is to reach 95% by the end of 1992.

The Diamond series of products are water-borne, have low VOC emissions, contain no ozone depletors, and are nonflammable.

The company is working with Environment Canada's Environmental Choice program to develop more categories and criteria for liquid films (i.e. stain, sealers, etc.). Diamond Colors carries the Ecologo for low toxic paint.

PRODUCT
Water-reduceable acrylic opaque lacquer

MANUFACTURER
Mills Paint and Wallcoverings
8380 River Rd.
Delta, BC V4G 1B5
Tel (604) 946-7011, (800) 663-6091   Fax (604) 946-0352

DESCRIPTION
ENVIROLAC 2100 PRIMER: Seals and forms an adhesive bond to the surface to increase the adhesion of subsequent top coats. Fast dry, sandable, hard finish.

Clean up with warm water when wet, solvent when dry.

ENVIROLAC 2200 SATIN, ENVIROLAC 2300 GLOSS: Fast drying, water-borne acrylic, low odor, nonflammable, easy cleanup, and a hard, nonyellowing, durable finish.

Colors: white bases are tintable to 4 oz. colorant. Deep and accent colors will not dry or develop final hardness as quickly as pastel colors due to the composition of the colorants.

Clean up with warm soapy water when wet, solvent when dry.

MANUFACTURER'S STATEMENT
Designed as a low VOC alternative to conventional white nitrocellulose lacquers (solvent-based fast dry enamels and lacquers) for interior surfaces.

PRODUCTION

PACKING SHIPPING

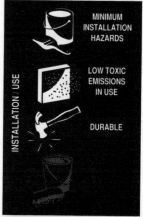

INSTALLATION / USE

MINIMUM
INSTALLATION
HAZARDS

LOW TOXIC
EMISSIONS
IN USE

DURABLE

RESOURCE RECOVERY

OTHER

MINIMUM
INSTALLATION
HAZARDS

LOW TOXIC
EMISSIONS
IN USE

DURABLE

PRODUCTION

PACKING / SHIPPING

INSTALLATION / USE

RESOURCE RECOVERY

OTHER

PRODUCT
Low toxicity water-based paints

MANUFACTURER
Pace Chem Industries, Inc.
779 La Grange Ave.
Newbury Park, CA 91320
Tel (805) 499-2911

RELATED PRODUCTS
RIGHT ON CRYSTAL AIRE CLEAR, p. 174

DESCRIPTION
CRYSTAL SHIELD LATEX PAINT: A pigmented form of Right On Crystal Shield, which provides the desired creaminess of texture and a degree of hiding power, with the ability to block out petrochemical, formaldehyde, and other fumes that are harmful to people with chemical sensitivities. Available in #1 Snowhite and #2 Antique White.

For use on walls (over existing paint surfaces) and other surfaces such as particle board, plywood, glue joints, linoleum, and floor tile.

MANUFACTURER'S STATEMENT
Crystal Shield's nontoxic formula blocks out toxic and irritating fumes (formaldehyde, petrochemicals, molds) from particle board, plywood, etc. Safe for children's furniture and food storage bins.

MANUFACTURER
Miller Paint Company
317 S.E. Grand Ave.
Portland, OR 97214
Tel (503) 233-4491

DESCRIPTION

6020 PRIMER SEALER: A high solids primer sealer for use on interior plaster, wallboard, brick, or concrete.

6450 INTERIOR VELVET FLAT LATEX: For use on walls and ceilings of drywall, cement block, and properly primed wood. Approx. 8% gloss.

1450 INTERIOR SATIN LATEX: For use on walls and ceilings. Will withstand repeated washing. Approx. 15% gloss.

2850 INTERIOR SEMI-GLOSS LATEX: For use on wall and trim, good for areas such as kitchens and bathrooms. Approx. 50% gloss.

7000 ACRYLIC EXTERIOR LATEX: Durable exterior paint, suitable for wood, concrete, masonry, and properly prepared metal. Approx. 15% gloss.

7054 ACRYLIC EXTERIOR SEMI-GLOSS LATEX ENAMEL: For wood siding and trim for doors, windows, gutters, etc. Approx. 60% gloss.

Products are white, but can be tinted with universal tinting color system.

MANUFACTURER'S STATEMENT

All low biocide, no fungicide paint products are from the regular product line, with the exception that they are manufactured without the addition of the biocide and fungicides that are usually added at the time of manufacture. This amounts to 90%–95% reduction of the biocide that would normally be in these paints. The 5%–10% that remains is in a premixed ingredient that is not manufactured by Miller Paint, but necessary in the manufacture of these products.

Miller Paint feels that these products will meet the needs of most hypersensitive individuals, though not all. They encourage trying a sample before using the product, and a free sample will be sent upon request.

PRODUCTION

PACKING / SHIPPING

MINIMUM INSTALLATION HAZARDS

LOW TOXIC EMISSIONS IN USE

DURABLE

INSTALLATION / USE

RESOURCE RECOVERY

OTHER

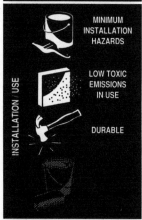

MINIMUM
INSTALLATION
HAZARDS

LOW TOXIC
EMISSIONS
IN USE

DURABLE

PRODUCT

Low fungicide, hypoallergenic water-based paints

MANUFACTURER

Murco Wall Products, Inc.
300 N.E. 21st St.
Fort Worth, TX 76106
Tel (817) 626-1987   Fax (817) 626-0821

DESCRIPTION

GF 1000: A premium quality, flat wall paint formulated without the slow releasing compounds and without a volatile fungicide. Its shelf life is controlled by a combination of pH control and in-can preservatives. Once the film is dry, the in-can preservative is permanently sealed in the film and does not become airborne or leach out. It is virtually odorless after being allowed to dry thoroughly and is comparable to conventional paints in color retention, scrubbability, and hiding powers.

LE 1000: A premium paint formulated similarly to GF 1000 except for a higher gloss. It is recommended for use where conventional latex enamels are used: kitchens, bathrooms, woodwork, etc. Because of the lack of a fungicide, mildew growth in bathrooms may increase with this paint.

MANUFACTURER'S STATEMENT

Murco's hypoallergenic products for the chemically sensitive offer non-asbestos formulation, low odor, high finished job quality, few application restrictions, and are cost comparable to conventional paints and joint compounds.

PRODUCT
Water-soluble wood dye and glaze

MANUFACTURER
Livos Plant Chemistry (Germany)
Imported by

Eco-House (1988) Inc.
P.O. Box 220, Stn. A
Fredericton, NB E3B 4Y9
Tel (506) 366-3529
Fax (506) 366-3577

Livos Plant Chemistry / USA
1365 Rufina Cir.
Santa Fe, NM 97214
   Tel (505) 438-3448

RELATED PRODUCTS
Livos oil finishes, p. 185
Livos shellacs, p. 168

DESCRIPTION
#750 BELA WOOD STAIN (WATER SOLUBLE): Extracts from dye plants, dissolved in water for use on interior untreated wood. Transparent dye which enhances the natural wood grain. Slightly photosensitive, and some alteration of color can occur if exposed to sunlight. Use of sealer/varnish is recommended (see above).

Colors: mignonette yellow, madder red, cochenille red, indigo blue, mistle green, yellow brown, walnut brown, cashew red-brown, black.

#272 TAYA VISCOUS WOOD GLAZE: Transparent, colored exterior and interior finish for all wood types in frequently wet area. Primer oil is recommended prior to use. Available clear, or in 3 premixed natural colors.

MANUFACTURER'S STATEMENT
See p. 147.

PRODUCTION

SUSTAINABLY ACQUIRED, OR RENEWABLE RESOURCE

LOW EMISSIONS PLANT

PACKING / SHIPPING

MINIMUM, RECYCLED, RECYCLABLE PACKAGING

INSTALLATION / USE

MINIMUM INSTALLATION HAZARDS

LOW TOXIC EMISSIONS IN USE

RESOURCE RECOVERY

OTHER

RESEARCH & EDUCATION PROGRAMS

PRODUCTION

SUSTAINABLY ACQUIRED, OR RENEWABLE RESOURCE

LOW EMISSIONS PLANT

PACKING / SHIPPING

MINIMUM, RECYCLED, RECYCLABLE PACKAGING

INSTALLATION / USE

MINIMUM INSTALLATION HAZARDS

LOW TOXIC EMISSIONS IN USE

RESOURCE RECOVERY

OTHER

RESEARCH & EDUCATION PROGRAMS

PRODUCT
All-natural staining paste

MANUFACTURER
Livos Plant Chemistry (Germany)
Imported by

Eco-House (1988) Inc.
P.O. Box 220, Stn. A
Fredericton, NB E3B 4Y9
Tel (506) 366-3529
Fax (506) 366-3577

Livos Plant Chemistry / USA
1365 Rufina Cir.
Santa Fe, NM 97214
Tel (505) 438-3448

RELATED PRODUCTS
#629 VINDO ENAMEL PAINT, p. 150
#221 DONNOS WEATHER PROTECTANT, p. 182
#222 CITRUS THINNER, p. 193
#315 GLEIVO LIQUID BEESWAX, p. 189
Livos oil finishes, p. 185

DESCRIPTION
#275 LEUKO STAINING PASTE (OIL BASED): Earthen and mineral staining pastes for most Livos oil-based products. Contains earthen and mineral pigments, linseed oil, chalk, talcum, silicic acid, #222 Citrus Thinner, and soya lecithin.

For exterior transparent finishes, only the darker pigments yield a strong enough protection from solar UV radiation. Interior colors: ochre, red ochre, ultramarine blue. Exterior and interior colors: terra di siena, English red, Persian red, rust brown, raw umber, burnt umber, black, green, white.

MANUFACTURER'S STATEMENT
See p. 147.

MANUFACTURER

Naturhaus (Germany)
Imported by

Lindquist Marketing
P.O. Box 3542
33118 Whidden Ave.
Mission, BC V2V 2T2
Tel (604) 826-1547

Lindquist Marketing
115 Garfield St.
Sumas, WA 98295
Tel (604) 826-1547
Fax (206) 988-2207

DESCRIPTION

#03000-110 WOOD GLAZE IN-EXTERIOR: Glaze with elastic and deep penetrating qualities. Dirt, ultraviolet, and water resistant.

Contains linseed oil, natural resins, orange turpentine, natural driers, and soil and oxide pigments. Available in 12 colors: neutral, pine, oak, mahogany, teak, walnut, rosewood, ebony, white, black, blue, and green.

For use on wood, cork, and stone for interior and exterior applications. Neutral glaze is for use in UV-protected areas only.

MANUFACTURER'S STATEMENT

See INSTALLATION MATERIALS, p. 118.

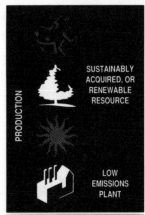

PRODUCTION

SUSTAINABLY ACQUIRED, OR RENEWABLE RESOURCE

LOW EMISSIONS PLANT

PACKING / SHIPPING

INSTALLATION USE

MINIMUM INSTALLATION HAZARDS

LOW TOXIC EMISSIONS IN USE

DURABLE

RESOURCE RECOVERY

OTHER

FAIR BUSINESS PRACTICES

PRODUCTION

LOW
EMISSIONS
PLANT

PACKING / SHIPPING

MINIMUM,
RECYCLED,
RECYCLABLE
PACKAGING

MINIMUM
INSTALLATION
HAZARDS

LOW TOXIC
EMISSIONS
IN USE

INSTALLATION / USE

RESOURCE RECOVERY

OTHER

RESEARCH &
EDUCATION
PROGRAMS

PRODUCT
Low toxicity water-based wood stains

MANUFACTURER
A.F.M. Enterprises Inc.
(American Formulating and Manufacturing)
1140 Stacy Court
Riverside, CA 92507
Tel (714) 781-6860, 781-6861   Fax (714) 781-6892

DESCRIPTION
DURO STAIN: Exterior and interior copolymer dispersion.

Comes in ready-to-use form in translucent shades of mahogany, maple, oak, walnut, birch, redwood, cedar, and clear.

Low odor, low toxicity, and noncombustible.

MANUFACTURER'S STATEMENT
See p. 152.

PRODUCT
Interior water-based wood stain

MANUFACTURER

Flecto Coatings
1000 45th St.
Oakland, CA 94608
Tel (510) 655-2470
Fax (510) 652-0969

Flecto Coatings
4260 Vanguard
Richmond, BC V6X 2P5
Tel (604) 273-8031
Fax (604) 273-6026

RELATED PRODUCTS

VARATHANE ELITE DIAMOND FINISH, p. 172

DESCRIPTION

DIAMOND "WOOD STAIN": Interior water-based semi-transparent acrylic stain. Available in 13 colors from mineral oxides.

MANUFACTURER'S STATEMENT

Flecto's ongoing environmental goal is to achieve zero discharge. To date, they have reduced hazardous wastes by 80%, and the target is to reach 95% by the end of 1992.

The Diamond series of products are water-borne, have low VOC emissions and contain no ozone depletors.

The company is working with Environment Canada's Environmental Choice program to develop more categories and criteria for liquid films (i.e. stain, sealers, etc.).

PRODUCTION

LOW
EMISSIONS
PLANT

PACKING / SHIPPING

MINIMUM
INSTALLATION
HAZARDS

LOW TOXIC
EMISSIONS
IN USE

INSTALLATION / USE

RESOURCE RECOVERY

OTHER

PRODUCTION

SUSTAINABLY ACQUIRED, OR RENEWABLE RESOURCE

LOW EMISSIONS PLANT

PACKING / SHIPPING

MINIMUM INSTALLATION HAZARDS

LOW TOXIC EMISSIONS IN USE

INSTALLATION / USE

RESOURCE RECOVERY

OTHER

RESEARCH & EDUCATION PROGRAMS

PRODUCT
All-natural resin lacquers

MANUFACTURER
Auro Organic Paints (Germany)
Imported by Sinan Co., Natural Building Materials
P.O. Box 857
Davis, CA 95617-0857
Tel (916) 753-3104

DESCRIPTION
NO. 211 CLEAR GLOSSY SHELLAC: Clear, quick drying lacquer. Suitable for insulating coat against penetrating water or tar/nicotine stains.

NO. 213 CLEAR VELVET SHELLAC: Clear, quick drying lacquer made from pure natural resin, dispersed in spirit, for semigloss coating of indoor wood surfaces. Ideal for application by spray gun.

NO. 215 CLEAR CEMBRA SHELLAC: Clear, quick drying lacquer for indoor wood, especially effective as protective inner coating against insects in cabinets, flour bins and mills.

NO. 222 CLEAR AMBER VARNISH: Clear genuine amber varnish for interior use on all types of wood, slight honey tone (especially suitable as the hardest natural finish for floor areas).

MANUFACTURER'S STATEMENT
See p. 145.

PRODUCT
All-natural wood varnishes

MANUFACTURER
Biofa Naturfarben GmbH (Germany)
Imported by Bau, Inc.
P.O. Box 190
Alton, NH 03839
Tel (603) 364-2400

RELATED PRODUCTS
PRIMING OIL #3750, p. 184
WOOD PRIMER WITH HERBS #1010, p. 184

DESCRIPTION
WOOD VARNISH—CLEAR #1075, WOOD VARNISH—COLORED #1061-1085: Produce highly elastic, silky matte finishes that repel water and dirt, but that allow the wood to breathe and are water-vapor permeable. Bind well to old and chemical paints. Varnishes contain linseed and other plant and ethereal oils, natural resins, clay and mineral pigments and natural solvents, without lead, cadmium or chromate. For interior and exterior use on smooth wooden surfaces such as walls, garage doors, planking, window frames, garden furniture, fences, ceilings, furniture, toys, etc. Available in 11 colors for interior and exterior use, and 6 colors for interior and protected exterior use only.

HARD VARNISH #2040 (GLOSSY), #2041(MATTE): Colorless varnish consisting of natural resins, linseed oil, other vegetable oils and natural solvents. Provides a tough and flexible finish, dirt and water repellent. For interior use on wood surfaces such as doors, furniture, wooden paneling, toys, cork, and stone. Especially suited for floors as an alternative to polyurethane.

MANUFACTURER'S STATEMENT
See p. 146.

PRODUCTION

SUSTAINABLY ACQUIRED, OR RENEWABLE RESOURCE

LOW EMISSIONS PLANT

PACKING / SHIPPING

INSTALLATION / USE

MINIMUM INSTALLATION HAZARDS

LOW TOXIC EMISSIONS IN USE

DURABLE

RESOURCE RECOVERY

OTHER

PRODUCTION

SUSTAINABLY ACQUIRED, OR RENEWABLE RESOURCE

LOW EMISSIONS PLANT

PACKING / SHIPPING

MINIMUM, RECYCLED, RECYCLABLE PACKAGING

INSTALLATION USE

MINIMUM INSTALLATION HAZARDS

LOW TOXIC EMISSIONS IN USE

RESOURCE RECOVERY

OTHER

RESEARCH & EDUCATION PROGRAMS

PRODUCT
All-natural alcohol-based shellacs and varnish

MANUFACTURER
Livos Plant Chemistry (Germany)
Imported by
Eco-House (1988) Inc.
P.O. Box 220, Stn. A
Fredericton, NB E3B 4Y9
Tel (506) 366-3529
Fax (506) 366-3577

Livos Plant Chemistry / USA
1365 Rufina Cir.
Santa Fe, NM 97214
Tel (505) 438-3448

RELATED PRODUCTS
#710 KIROS ALCOHOL THINNER, p. 193

#711 TEKIS FLATTENING AGENT: Used to regulate the gloss of all alcohol-based Livos shellac products. For interior use only.

DESCRIPTION
#701 LANDIS CLEAR SHELLAC: Clear shellac, highly suitable for transparent, high gloss finishing of fine furniture, closets and cabinets, doors, wainscoting. Not for use on areas exposed to moisture or alcohol, such as floors, counter tops, etc. Safe for use on toys and other children's items. Seals up to 80% of formaldehyde emissions from particle board and plywood. Qualifies for French polishing.

#709 TREBO NATURAL SHELLAC: Warm, semi-shiny, honey-toned shellac, for finishing of fine furniture, doors, and wainscoting. Not for use on areas exposed to moisture or alcohol, such as floors, counter tops, etc. Seals up to 80% of formaldehyde emissions from particle board and plywood. Sheen can be flattened by adding #711 Tekis flattening agent.

#602 TUNNA FLOOR & FURNITURE VARNISH: Heavy body, transparent sealing varnish for interior use. For hardwood floors, doors, and furniture.

MANUFACTURER'S STATEMENT
See p. 147.

PRODUCT
All-natural lacquer finishes

MANUFACTURER

Naturhaus (Germany)
Imported by
Lindquist Marketing
P.O. Box 3542
33118 Whidden Ave.
Mission, BC V2V 2T2
Tel (604) 826-1547

Lindquist Marketing
115 Garfield St.
Sumas, WA 98295
Tel (604) 826-1547
Fax (206) 988-2207

DESCRIPTION

#04000 VARNISH / LACQUER (INTERIOR): Waterproof and durable sealer, that highlights the grain. Neutral color. Contains linseed oil, wood oil, natural resin, lime resin, orange turpentine, and natural driers. For use with wood, cork, tiles, and stone surfaces. For interior use, but not for floor applications.

#04050 WOOD VARNISH / LACQUER (EXTERIOR): Elastic and waterproof sealer that allows the wood to react to outside weather conditions. Neutral color. Highlights the grain, but is not ultraviolet resistant. Contains linseed oil, wood oil, natural resin, orange turpentine, and natural driers.

#04100 FLOOR LACQUER / VARNISH (INTERIOR): Durable, water-resistant floor sealer, that highlights the grain. Contains linseed oil, wood oil, natural resin, lime resin, orange turpentine, and natural driers. For use on wood stairs and working areas covered with cork, wood, or stone. For interior use only.

MANUFACTURER'S STATEMENT

See INSTALLATION MATERIALS, p. 118.

PRODUCTION

SUSTAINABLY ACQUIRED, OR RENEWABLE RESOURCE

LOW EMISSIONS PLANT

PACKING / SHIPPING

INSTALLATION / USE

MINIMUM INSTALLATION HAZARDS

LOW TOXIC EMISSIONS IN USE

DURABLE

RESOURCE RECOVERY

OTHER

MINIMUM
INSTALLATION
HAZARDS

LOW TOXIC
EMISSIONS
IN USE

DURABLE

SIMPLE,
NONTOXIC
MAINTENANCE

PRODUCT
Low toxicity interior clear varnishes/sealers

MANUFACTURER
BonaKemi USA, Inc.
14805 E. Moncrieff Pl.
Aurora, CO 80011
Tel (303) 371-1411, (800) 872-5515   Fax (303) 371-6958

DESCRIPTION
All products in the *Pacific Swedish Finishes Line* provide good adhesion, resistance to most household chemicals, are nonambering, voc compliant, have a high solid content and are nonflammable. For interior use only.

HIGH BUILD SEALER: Formulated to give hardwood floors high build, depth of finish, and long wear. For use on natural, stained, or newly sanded hardwood floors. Easy to apply, quick drying, and specially formulated to reduce grain raise. For use with any BonaKemi Waterborne protective top coat finishes.

PACIFIC ONE: Pacific One Waterborne Finish is a 1-component product on a polyurethane/acrylic water dispersion base that is easy to use and requires no mixing. Specially formulated for light traffic on all types of residential hardwood floors. Available in gloss or silkmat sheen.

PACIFIC II: Pacific II is a 2-component, nonyellowing waterborne finish for residential wood floors. Formulated for normal wear areas.

PACIFIC SPORT: 3-product system includes sealer, finish, and cleaner concentrate. For use on sports floors, to keep a light airy look for better overall lighting effect. Pacific Sport offers a high gloss finish for sports floors, and a satin sheen for stages and dance floors.

PACIFIC STRONG: Pacific Strong Waterborne Finish is a 2-component product on a polyurethane/acrylic dispersion base, formulated for use on high traffic hardwood floors, such as restaurants, hotels, museums, sports arenas, and residences. Provides a tough, durable, and long lasting surface protection. Available in gloss or silkmat sheen.

WOODLINE ULTRACURE: A waterborne wood floor finish that reproduces the rich color and performance of solvent-based urethanes. Provides a strong, long-lasting chemical and scuff resistant wear layer. Reduces grain raise, while filling and bridging the grain of wood. Formulated to be virtually odor free, nonflammable, and for water cleanup. Not recommended on white or pastel floors. Available in gloss, semigloss, and satin sheens.

MANUFACTURER'S STATEMENT
BonaKemi has a strong recycling program, as part of its commitment to the environment, including aluminum, and bond and computer paper. Their newsletter is printed on recycled paper.

MANUFACTURER

Environmental Technology, Inc.
S. Bay Depot Rd.
Fields Landing, CA 95537
Tel (707) 443-9323
Fax (707) 443-7962

Environmental Technology, Inc. Canada
Unit 305, 2056 #10 Highway
Langley, BC V3A 6K8
Tel (604) 534-3810
Fax (604) 534-4173

DESCRIPTION

ENVIROSPRAY POLYURETHANE CLEARS: Recommended for coating over interior and exterior wood surfaces. Available in gloss and satin sheens.

ENVIROSPRAY ACRYLIC CLEARS: Recommended for all other surfaces, as well as wood. Available in gloss and satin sheens.

Excellent 1-coat coverage on nonporous surfaces, reduced overspray, soap and water cleanup, salt and weather resistant, will not crack or peel, high fading resistance.

Formulations contain nonchlorinated solvents that are virtually nontoxic. Dimethyl ether is used as a propellant (ozone safe); no CFCs or freon are used. Nonflammable.

Long shelf life due to water-based system.

MANUFACTURER'S STATEMENT

ENVIROSpray is the only spray paint to be classified with the Blue Angel Symbol, of the European Common Market, designated as an environmentally friendly product.

Contains no methylene chloride solvent, lead, ketones, hydrocarbons, aromatics, or other petroleum distillates.

PRODUCTION

PACKING / SHIPPING

INSTALLATION / USE

MINIMUM INSTALLATION HAZARDS

LOW TOXIC EMISSIONS IN USE

DURABLE

SIMPLE, NONTOXIC MAINTENANCE

RESOURCE RECOVERY

OTHER

PRODUCT
Low toxicity interior clear varnish/sealer

MANUFACTURER

Flecto Coatings
4260 Vanguard
Richmond, BC V6X 2P5
Tel (604) 273-8031
Fax (604) 273-6026

Flecto Coatings
1000 45th St.
Oakland, CA 94608
Tel (510) 655-2470
Fax (510) 652-0969

DESCRIPTION

VARATHANE ELITE DIAMOND FINISH : Interior, clear sealer, in an "interpenetrating polymer network," a blend of acrylic and polyurethane. Finish is water-clear, and imparts hardness and abrasion resistance. Water cleanup.

Available in matte, semigloss, and gloss sheens.

MANUFACTURER'S STATEMENT

See p. 165.

PRODUCTION

LOW
EMISSIONS
PLANT

PACKING / SHIPPING

INSTALLATION / USE

MINIMUM
INSTALLATION
HAZARDS

LOW TOXIC
EMISSIONS
IN USE

DURABLE

SIMPLE,
NONTOXIC
MAINTENANCE

RESOURCE RECOVERY

OTHER

PRODUCT
Low toxicity clear sealer finishes

MANUFACTURER
A.F.M. Enterprises Inc.
(American Formulating and Manufacturing)
1140 Stacy Court
Riverside, CA 92507
Tel (714) 781-6860, 781-6861   Fax (714) 781-6892

RELATED PRODUCTS
STRIPPER 66, p. 196
SUPER CLEAN, p. 200

DESCRIPTION
DYNO SEAL: Liquid copolymer that forms a durable waterproof membrane for an all-weather resistant coating, sealer, and adhesive. For use below-grade or as a roof and deck coating.

HARD SEAL: Medium gloss for floors, cabinets, woodwork, porous tile, or concrete. Will seal out obnoxious and toxic fumes.

KLEAR SEAL: Acrylated copolymers in water base for use as interior and exterior sealer. For metal, masonry, and concrete.

PENETRATING WATER SEAL: Above-grade water repellent for use on porous bricks, Mexican-type pavers, concrete blocks, stucco, concrete, and raw wood. Reduces water absorption, controls effloresence.

POLYURASEAL: Durable, low gloss, mar-resistant polyester-urethane polymer in water-based emulsion. For use on wood, metal, vinyl-coated fabrics, and other plastic surfaces.

WATER SEAL: Interior and exterior sealer to reduce the emissions of existing finishes. May be used as sealer on grout and porous surfaces, such as quarry and unglazed tile, terrazzo, etc.

ACRYLACQ: Clear, high gloss, water-based, lacquer-like finish for wood, vinyl, and certain types of ceramic.

MANUFACTURER'S STATEMENT
See p. 152.

PRODUCTION

LOW EMISSIONS PLANT

PACKING / SHIPPING

MINIMUM, RECYCLED, RECYCLABLE PACKAGING

INSTALLATION / USE

MINIMUM INSTALLATION HAZARDS

LOW TOXIC EMISSIONS IN USE

DURABLE

SIMPLE, NONTOXIC MAINTENANCE

RESOURCE RECOVERY

OTHER

RESEARCH & EDUCATION PROGRAMS

PRODUCTION

PACKING / SHIPPING

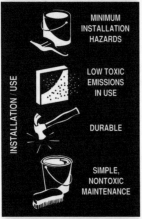

INSTALLATION / USE

MINIMUM
INSTALLATION
HAZARDS

LOW TOXIC
EMISSIONS
IN USE

DURABLE

SIMPLE,
NONTOXIC
MAINTENANCE

RESOURCE RECOVERY

OTHER

PRODUCT
Hypoallergenic clear sealer finish

MANUFACTURER
Pace Chem Industries, Inc.
779 La Grange Ave.
Newbury Park, CA 91320
Tel (805) 499-2911

DESCRIPTION
RIGHT-ON CRYSTAL AIRE CLEAR FINISH: Water-borne acrylic clear sealer that is nontoxic and hypoallergenic. May be used on most nonwaxy surfaces, including: cabinets, woodwork, wood furniture, paneling, plasterboard, colorfast wallpaper, vinyl and clay tile, cardboard, fabrics, latex and oil paints, urethanes and lacquer, glass, and all metals.

Water clear (will never yellow) and remains flexible. Scrubbable and durable substitute for lacquer, varnish, and urethane. Resists scratching, peeling and chipping. Water cleanup.

Available in matte, satin, and gloss sheens.

MANUFACTURER'S STATEMENT
Crystal Shield's nontoxic formula blocks out toxic and irritating fumes (formaldehyde, petrochemicals, molds) from particle board, plywood, etc. Safe for children's furniture and food storage bins.

MANUFACTURER

A.F.M. Enterprises Inc.
(American Formulating and Manufacturing)
1140 Stacy Court
Riverside, CA 92507
Tel (714) 781-6860, 781-6861   Fax (714) 781-6892

DESCRIPTION

CEM BOND CONCRETE AND MASONRY PAINT: Water-based emulsion multi-purpose paint to coat and seal concrete and masonry surfaces. May also be used for exterior or interior wood trim.

May be applied over new and previously painted surfaces; will give good performance with only one coat.

Cem Bond comes in white and gray, and may be tinted to any light to medium shade with all universal water-based tinting systems.

MANUFACTURER'S STATEMENT

See p. 152.

PRODUCTION

LOW
EMISSIONS
PLANT

PACKING / SHIPPING

MINIMUM,
RECYCLED,
RECYCLABLE
PACKAGING

INSTALLATION / USE

MINIMUM
INSTALLATION
HAZARDS

LOW TOXIC
EMISSIONS
IN USE

DURABLE

RESOURCE RECOVERY

OTHER

RESEARCH &
EDUCATION
PROGRAMS

PRODUCTION

PACKING / SHIPPING

INSTALLATION / USE

DURABLE

SIMPLE,
NONTOXIC
MAINTENANCE

RESOURCE RECOVERY

OTHER

PRODUCT
Acrylic flooring sealer

MANUFACTURER

General Polymers
145 Caldwell Dr.
Cincinnati, OH 45216
Tel (513) 761-0011
Fax (513) 761-4496

General Polymers Canada
405 The West Mall, Suite 700
Etobicoke, ON M9C 5J1
Tel (416) 621-3988

DESCRIPTION

MACROSEPTIC® ACRYLIC SEALER: Acrylic sealer, with Intersept® to inhibit the growth of harmful bacteria and fungi.

Suited for working environments that demand high performance and hygienic freshness, where the floor is not in need of replacement.

HIGH-PERFORMANCE SEALER: Clear acrylic sealer designed for use over terrazzo, tile grout, bare concrete, and a variety of other substrates.

MANUFACTURER'S STATEMENT

General Polymers has been developing composition resin flooring and coating systems to solve the challenges presented by specialized environments. Chemists and technicians use state-of-the-art polymer technology in research and development to meet the broadest array of flooring needs.

General Polymers is a partner of the Envirosense™ consortium who provide products manufactured with Intersept®, a broad spectrum biocide.

NOTE

The addition of biocides to materials may not be an appropriate practice for many applications. Biocides are not a substitute for adequate cleaning and maintenance.

PRODUCT
Low toxicity concrete sealers

MANUFACTURER
L. M. Scofield Company
6533 Bandini Blvd.
Los Angeles, CA 90040-3182
Tel (213) 723-5285
Fax (213) 722-6029

DESCRIPTION
LITHOTHANE CONCRETE SEALANT: Clear urethane coating. Excellent adhesion to concrete and resistance to chemicals, weathering, and graffiti.

REPELLO PROTECTIVE SURFACE COATING: Transparent, nontoxic, water-based, sacrificial coating to protect vertical surfaces from graffiti and other contaminants.

CEMENTONE CLEAR SEALER: Ready-to-use, water-based, clear, semigloss sealer for antiqued, imprinted, multicolored, and uncolored concrete flatwork. Durable and complies to the most stringent air quality regulations.

COLORCURE CONCRETE SEALER: Ready-to-use, water-based, color-matched curing and finishing system for freshly placed concrete, and a durable sealer for maintaining colored or uncolored concrete flatwork. Durable and complies to the most stringent air quality regulations.

MANUFACTURER'S STATEMENT
The vast majority of L. M. Scofield's products are water based and low odor. A company policy has been adopted so that all new product development, and discontinued products which contain more toxic substances (i.e. heavy metals), will be replaced with only low toxicity, water-based formulations.

PRODUCTION

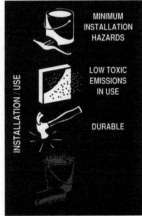

PACKING / SHIPPING

INSTALLATION / USE

MINIMUM INSTALLATION HAZARDS

LOW TOXIC EMISSIONS IN USE

DURABLE

RESOURCE RECOVERY

OTHER

PRODUCTION

LOW
EMISSIONS
PLANT

PACKING / SHIPPING

MINIMUM,
RECYCLED,
RECYCLABLE
PACKAGING

INSTALLATION / USE

MINIMUM
INSTALLATION
HAZARDS

LOW TOXIC
EMISSIONS
IN USE

RESOURCE RECOVERY

OTHER

RESEARCH &
EDUCATION
PROGRAMS

PRODUCT
Low toxicity flame retardant finish for wood

MANUFACTURER
A.F.M. Enterprises Inc.
(American Formulating and Manufacturing)
1140 Stacy Court
Riverside, CA 92507
Tel (714) 781-6860, 781-6861   Fax (714) 781-6892

DESCRIPTION
   SHINGLE PROTEK: Resinous penetrating fire retardant treatment for use on wood-shingled roof tops (already in place). Resistant to water leaching, and protects against spontaneous fires. Nonflammable and nonhazardous.

MANUFACTURER'S STATEMENT
See p. 152.

Low toxicity water repellent finish for carpet

MANUFACTURER

A.F.M. Enterprises Inc.
(American Formulating and Manufacturing)
1140 Stacy Court
Riverside, CA 92507
Tel (714) 781-6860, 781-6861   Fax (714) 781-6892

DESCRIPTION

CARPET GUARD: Nonflammable, air drying/air curing, water-soluble siliconate. Forms an insoluble water resistant film, and odor and bacteria barrier. Repels waterborne stains and sheds water easily.

For use on synthetic fibers only, not for use on wool carpets.

MANUFACTURER'S STATEMENT

See p. 152.

NOTE

Sealing carpet fibers is not a substitute for adequate cleaning and maintenance. Carpeting may not be an appropriate floor covering for heavy soil areas.

PRODUCTION

LOW EMISSIONS PLANT

PACKING / SHIPPING

MINIMUM, RECYCLED, RECYCLABLE PACKAGING

INSTALLATION / USE

MINIMUM INSTALLATION HAZARDS

LOW TOXIC EMISSIONS IN USE

RESOURCE RECOVERY

OTHER

RESEARCH & EDUCATION PROGRAMS

LOW TOXIC
EMISSIONS
IN USE

DURABLE

PRODUCT
Low toxicity concrete masonry water repellent treatment

MANUFACTURER
L. M. Scofield Company
6533 Bandini Blvd.
Los Angeles, CA 90040-3182
Tel (213) 723-5285
Fax (213) 722-6029

DESCRIPTION
REPELLO WATER REPELLENT: Penetrating, water-based water repellent used to protect vertical and horizontal concrete and masonry surfaces from damage due to water and water-borne contaminants.

MANUFACTURER'S STATEMENT
See p. 177.

PRODUCT
Low toxicity antimicrobial agent

MANUFACTURER
A.F.M. Enterprises Inc.
(American Formulating and Manufacturing)
1140 Stacy Court
Riverside, CA 92507
Tel (714) 781-6860, 781-6861   Fax (714) 781-6892

RELATED PRODUCTS
SUPER CLEAN, p. 200

DESCRIPTION
X 158 MILDEW CONTROL: Safe, high potency liquid blended with surfactants that provide mildew resistance and preservation against microbial and bacterial attack. Will not remove existing mildew growth, but will control continued growth.

For use in tub enclosures, shower stalls, bathroom walls, roof tops, etc.

MANUFACTURER'S STATEMENT
See p. 152.

NOTE
The use of biocides is not a substitute for adequate cleaning and maintenance.

PRODUCTION

LOW
EMISSIONS
PLANT

PACKING / SHIPPING

MINIMUM,
RECYCLED,
RECYCLABLE
PACKAGING

INSTALLATION / USE

MINIMUM
INSTALLATION
HAZARDS

LOW TOXIC
EMISSIONS
IN USE

RESOURCE RECOVERY

OTHER

RESEARCH &
EDUCATION
PROGRAMS

PRODUCTION

SUSTAINABLY ACQUIRED, OR RENEWABLE RESOURCE

LOW EMISSIONS PLANT

PACKING / SHIPPING

MINIMUM, RECYCLED, RECYCLABLE PACKAGING

INSTALLATION / USE

MINIMUM INSTALLATION HAZARDS

LOW TOXIC EMISSIONS IN USE

DURABLE

SIMPLE, NONTOXIC MAINTENANCE

RESOURCE RECOVERY

OTHER

RESEARCH & EDUCATION PROGRAMS

PRODUCT
All-natural wood preservers

MANUFACTURER
Livos Plant Chemistry (Germany)
Imported by
Eco-House (1988) Inc.
P.O. Box 220, Stn. A
Fredericton, NB E3B 4Y9
Tel (506) 366-3529
Fax (506) 366-3577

Livos Plant Chemistry / USA
1365 Rufina Cir.
Santa Fe, NM 97214
Tel (505) 438-3448

RELATED PRODUCTS
#220 CITRUS THINNER, p. 193
#270 KALDET RESIN AND OIL FINISH, p. 185
#275 LEUKO STAINING PASTES, p. 162

DESCRIPTION
#221 DONNOS WEATHER PROTECTANT: Used as an alternative to pressure-treated lumber for fences, decks, flower boxes, playground equipment, barns, and other outdoor structures.

Water repellent, weather (ultraviolet) resistant, and stays permanently elastic. Easy refinishing by application of additional coats. Sealer coat is recommended for items which are touched often.

#256 ANDRASTOS BORON SALT: Water-soluble boron salt wood preserver. Protects against fungi, mildew, and insects without changing color of the wood. Can be used in areas such as food storage shelves, where other preservatives are not acceptable. Use on interior surfaces, partly covered exterior wood, or in conjunction with an exterior finishing product.

MANUFACTURER'S STATEMENT
See p. 147.

PRODUCT
All-natural wood preserver

MANUFACTURER
Naturhaus (Germany)
Imported by

Lindquist Marketing
P.O. Box 3542
33118 Whidden Ave.
Mission, BC V2V 2T2
Tel (604) 826-1547

Lindquist Marketing
115 Garfield St.
Sumas, WA 98295
Tel (604) 826-1547
Fax (206) 988-2207

RELATED PRODUCTS
Naturhaus oil products, p. 186
Naturhaus lacquers, p. 169
#03000-110 WOOD GLAZE, p. 163
Naturhaus waxes, p. 190

DESCRIPTION
BORAX WOOD PROTECTION: Pesticidal and fungicidal qualities for wood preservation. Interior and exterior wood protection for areas not in contact with soil or water. Neutral color. Contains borax and wood resin granules.

MANUFACTURER'S STATEMENT
See INSTALLATION MATERIALS, p. 118.

PRODUCTION

SUSTAINABLY ACQUIRED, OR RENEWABLE RESOURCE

LOW EMISSIONS PLANT

PACKING / SHIPPING

MINIMUM INSTALLATION HAZARDS

LOW TOXIC EMISSIONS IN USE

INSTALLATION / USE

RESOURCE RECOVERY

OTHER

FAIR BUSINESS PRACTICES

SUSTAINABLY
ACQUIRED, OR
RENEWABLE
RESOURCE

LOW
EMISSIONS
PLANT

PRODUCTION

PACKING / SHIPPING

MINIMUM
INSTALLATION
HAZARDS

LOW TOXIC
EMISSIONS
IN USE

INSTALLATION / USE

RESOURCE RECOVERY

OTHER

PRODUCT
All-natural oil finishes

MANUFACTURER
Biofa Naturfarben GmbH (Germany)
Imported by Bau, Inc.
P.O. Box 190
Alton, NH 03839
Tel (603) 364-2400

RELATED PRODUCTS
Biofa wood varnishes, p. 167
Biofa waxes, p. 188

DESCRIPTION
WOOD PRIMER WITH HERBS #1010: Transparent, water-resistant, oil-based primer prepared from plant extracts, ethereal oils, vegetable oils, and natural resins in natural solvents. Helps protect wood against pest and insect infestations. For interior and exterior use on shingles, windows, garden furniture, decks, cold frames, play structures, humid rooms.

PRIMING OIL #3750: Water-repelling wood priming oil prepared from quality lead-free linseed oil and natural solvents. Penetrates deep into the wood's fibers, bringing out the natural grain structure. For interior use on absorbent surfaces, such as wood, cork, stone, open-pored tiles, etc.

HARD OIL #2045: Penetrates deep into the wood, revitalizes its natural structure and enhances the natural tones. Surfaces are open pored, silky matte, and water repellent. Contains high quality plant oils, tree resins, and natural solvents. For interior use on absorbent surfaces, such as wood furniture, paneling and floors, cork, and unglazed tile.

MANUFACTURER'S STATEMENT
See PAINTS, p. 146.

PRODUCT
All-natural oil and resin finishes for furniture and floors

MANUFACTURER
Livos Plant Chemistry (Germany)
Imported by

Eco-House (1988) Inc.
P.O. Box 220, Stn. A
Fredericton, NB E3B 4Y9
Tel (506) 366-3529
Fax (506) 366-3577

Livos Plant Chemistry / USA
1365 Rufina Cir.
Santa Fe, NM 97214
Tel (505) 438-3448

RELATED PRODUCTS
#222 CITRUS THINNER, p. 193
#275 LEUKO STAINING PASTE, p. 162

DESCRIPTION
#270 KALDET RESIN AND OIL FINISH: Versatile wood finishing product, for interior and exterior use. Penetrates deep and hardens the surface. Water repellent, and permanently elastic, allows the wood to breathe and exchange moisture. Won't blister or peel. Available in clear, and 12 premixed colors.

#261 DUBNO PRIMER OIL, #260 LINUS LINSEED VARNISH: Economical priming treatment prior to treatment with oil and resin finishes or beeswax products. Can be used alone as a dirt repelling indoor finish on low use wood surfaces which are not frequently exposed to moisture. Intensifies the wood grain and creates a satin surface sheen. #260 Linus linseed varnish can be used on cedar, cork, porous tiles, or porous natural stone.

#264 MELDOS HARD SEALER OIL FOR SOFT WOODS: Impregnates wood and becomes water resistant. For use on interior floors, stairs, and other high use areas of softwood, cork, porous stone, terracotta tiles, concrete, and brick. Water repellent and resistant. Will not blister or peel. Intensifies the natural wood grain. Safe for treating toys, cribs, kitchen counters, cutting boards, and other wooden kitchen utensils. Honey-colored, but can be tinted.

#266 ARDVOS HARDWOOD FLOOR OIL: For interior use on any type of hardwood floor and furniture. Clear finish that functions like a penetrating oil and a film-type finish. Not for use on softwood or cork. Water repellent and allows the surface to breathe. Satin finish.

MANUFACTURER'S STATEMENT
See p. 147.

PRODUCTION

SUSTAINABLY ACQUIRED, OR RENEWABLE RESOURCE

LOW EMISSIONS PLANT

PACKING / SHIPPING

MINIMUM, RECYCLED, RECYCLABLE PACKAGING

INSTALLATION / USE

MINIMUM INSTALLATION HAZARDS

LOW TOXIC EMISSIONS IN USE

RESOURCE RECOVERY

OTHER

RESEARCH & EDUCATION PROGRAMS

PRODUCTION

SUSTAINABLY
ACQUIRED, OR
RENEWABLE
RESOURCE

LOW
EMISSIONS
PLANT

PACKING / SHIPPING

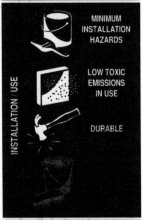

INSTALLATION / USE

MINIMUM
INSTALLATION
HAZARDS

LOW TOXIC
EMISSIONS
IN USE

DURABLE

RESOURCE RECOVERY

OTHER

PRODUCT
All-natural oil finishes

MANUFACTURER
Naturhaus (Germany)
Imported by

Lindquist Marketing	Lindquist Marketing
P.O. Box 3542	115 Garfield St.
33118 Whidden Ave.	Sumas, WA 98295
Mission, BC V2V 2T2	Tel (604) 826-1547
Tel (604) 826-1547	Fax (206) 988-2207

RELATED PRODUCTS
#03000-110 WOOD GLAZE, p. 163
Naturhaus lacquers, p. 169
Naturhaus waxes, p. 190

DESCRIPTION
NATURAL RESIN OIL—EXTERIOR AND INTERIOR USE: Porous, clear oil designed to absorb moisture. Moisture and dirt resistant, highlights the grain and penetrates deeply into the wood. Contains linseed oil, wood oil, resin and pine oils, orange turpentine, and natural driers. For impregnating wood, cork, and stone surfaces.

#00210 NATURAL RESIN HARDENING OIL: Porous, light oil designed to seal wood surfaces and make them water resistant. Highlights the grain and makes a very durable wood surface. Contains linseed oil, wood oil, resin and pine oils, orange turpentine, and natural driers. For impregnating wood, cork, and stone surfaces, and surface protection in general.

#00100 LINSEED OIL: Porous, clear oil designed to absorb moisture. Highlights the grain with a honey tone. Stays pliable and stable to 212°F (100°C). Contains linseed oil and natural driers. For impregnating all porous surfaces such as wood, cork, tiles, and stone surfaces. For interior use only.

MANUFACTURER'S STATEMENT
See INSTALLATION MATERIALS, p. 118.

MANUFACTURER
Auro Organic Paints (Germany)
Imported by Sinan Co., Natural Building Materials
P.O. Box 857
Davis, CA 95617-0857
Tel (916) 753-3104

DESCRIPTION

NO. 171 FLOOR AND FURNITURE PLANT WAX: Ointment of beeswax and plain waxes for water-repellent upgrading of wooden surfaces, especially for floors. Provides velvety surface.

NO. 173 LARCH RESIN FURNITURE WAX: Transparent beeswax paste wax with larchwood resin for indoor use on wood, cork, and smooth natural stones.

NO. 181 LIQUID BEESWAX FINISH: Liquid beeswax for water-repellent surface upgrading of wood, cork, clay tiles, etc. Easy to apply by brush or spray gun, to be buffed after short drying time. Allows surface to breathe, controls static buildup.

NO. 421 WAX CLEANER: Biological cleaner for soiled, wax-treated surfaces, slightly rewaxing.

NO. 431 BEESWAX FLOOR CARE: Organic emulsion for regular care of waxed floors, linoleum, etc. as an additive for washing water. Forms water-repellent protective coating.

MANUFACTURER'S STATEMENT
See p. 145.

PRODUCTION

SUSTAINABLY ACQUIRED, OR RENEWABLE RESOURCE

LOW EMISSIONS PLANT

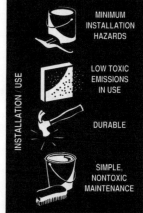

PACKING / SHIPPING

INSTALLATION / USE

MINIMUM INSTALLATION HAZARDS

LOW TOXIC EMISSIONS IN USE

DURABLE

SIMPLE, NONTOXIC MAINTENANCE

RESOURCE RECOVERY

OTHER

RESEARCH & EDUCATION PROGRAMS

SUSTAINABLY
ACQUIRED, OR
RENEWABLE
RESOURCE

LOW
EMISSIONS
PLANT

PRODUCTION

PACKING / SHIPPING

MINIMUM
INSTALLATION
HAZARDS

LOW TOXIC
EMISSIONS
IN USE

INSTALLATION / USE

RESOURCE RECOVERY

OTHER

PRODUCT
All-natural wax finishes

MANUFACTURER
Biofa Naturfarben GmbH (Germany)
Imported by Bau, Inc.
P.O. Box 190
Alton, NH 03839
Tel (603) 364-2400

DESCRIPTION
HARD FLOOR WAX #2060: Clear, silky matte, hard sealing floor wax prepared from bee and plant waxes, plant extracts, balsam, ethereal oils, and natural solvents. May be used to protect all soft and hard woods. For use on wood and clay tile floors, stairs, paneling, nonglazed stone, and cork.

LIQUID WAX #2075: Fragrant, transparent, glossy cleaning polish from beeswax, mineral waxes, and natural solvents. Sprayable and easy to use, penetrates the wood, producing an antistatic, dirt and water resistant surface. For interior use on unlacquered, waxed, stained, or oiled smooth woods. Suitable for toys, furniture, paneling, etc.

BALSAM BEESWAX #2070: Fragrant, water-repellent wax prepared from beeswax and other natural waxes, plant and ethereal oils, and natural solvents. Penetrates deep to maintain, protect, and enhance the wood. Breathes and is antistatic. Suitable for all types of smooth interior wood surfaces. May be used on waxed, stained, or oiled surfaces, such as toys, furniture, paneling, antiques, musical instruments, etc.

MANUFACTURER'S STATEMENT
See p. 146.

PRODUCT
All-natural furniture polishes and waxes

MANUFACTURER

Livos Plant Chemistry (Germany)
Imported by
Eco-House (1988) Inc.
P.O. Box 220, Stn. A
Fredericton, NB E3B 4Y9
Tel (506) 366-3529
Fax (506) 366-3577

Livos Plant Chemistry / USA
1365 Rufina Cir.
Santa Fe, NM 97214
Tel (505) 438-3448

RELATED PRODUCTS

Livos shellacs, p. 168
Livos oil finishes, p. 185

DESCRIPTION

 # 560 DRYADEN FURNITURE POLISH: For maintenance and cleaning of stone, cork, or wood surfaces, with oil-based finishes and shellacs.

#564 ARVEN POLISH: Contains essential oil of the cembra pine, a known natural insect repellent. Can be used as a cleaner and moth repellent.

#565 ALIS FURNITURE POLISH: For use on surfaces treated with hard sealer oil. Removes stains and improves scratch resistance.

#303 BILO FLOOR WAX: Protectant finish for materials such as wood, stone, brick, terracotta. Can also be used for furniture. Not for use on materials that are frequently wet.

#312 BEKOS FURNITURE WAX: Protectant finish for furniture and other materials such as wood, cork, linoleum. Forms nonstatic, breathable film. Not for use on materials that are frequently wet.

#315 GLEIVO LIQUID BEESWAX: Interior surface finishing and care wax with moderate cleaning effect and protection. Not for use on materials that are frequently wet. May be stained. Sprayable.

MANUFACTURER'S STATEMENT

See p. 147.

PRODUCTION

SUSTAINABLY ACQUIRED, OR RENEWABLE RESOURCE

LOW EMISSIONS PLANT

PACKING / SHIPPING

MINIMUM, RECYCLED, RECYCLABLE PACKAGING

INSTALLATION / USE

MINIMUM INSTALLATION HAZARDS

LOW TOXIC EMISSIONS IN USE

SIMPLE, NONTOXIC MAINTENANCE

RESOURCE RECOVERY

OTHER

RESEARCH & EDUCATION PROGRAMS

SUSTAINABLY ACQUIRED, OR RENEWABLE RESOURCE

LOW EMISSIONS PLANT

PRODUCTION

PACKING / SHIPPING

MINIMUM INSTALLATION HAZARDS

LOW TOXIC EMISSIONS IN USE

INSTALLATION / USE

RESOURCE RECOVERY

FAIR BUSINESS PRACTICES

OTHER

PRODUCT
Miscellaneous all-natural wax products

MANUFACTURER
Naturhaus (Germany)
Imported by

Lindquist Marketing
P.O. Box 3542
33118 Whidden Ave.
Mission, BC V2V 2T2
Tel (604) 826-1547

Lindquist Marketing
115 Garfield St.
Sumas, WA 98295
Tel (604) 826-1547
Fax (206) 988-2207

RELATED PRODUCTS
NATURAL THINNER #1, p. 194
#03000-110 WOOD GLAZE, p. 163
Naturhaus lacquers, p. 169

DESCRIPTION
#01030 BEESWAX: Matte protective coating for wood. Protects wood against water, soil, and dust. Has antistatic properties. Contains linseed oil, beeswax, carnauba wax, resin, and natural driers. Good surface protection for wood, cork, and natural stone surfaces exposed to moderate to heavy wear. Use after wax-impregnating lacquers and glazes.

#01010 INTERIOR FINISHING WAX, #01000 LIQUID BEESWAX: Transparent matte finishing wax is antistatic, porous, water and soil resistant. Contains linseed oil, resin oil, natural resin, beeswax, carnauba wax, orange turpentine and natural driers. Topical protection for untreated, light wear surfaces of wood, cork, and natural stone.

#01050 IMPREGNATING WAX: Matte protective impregnating protection without thinners. Heat resistant up to 250°F (120°C). Use for impregnating surfaces of wood, cork, stone, and tile in high traffic areas. Contains linseed oil, wood oil, natural resin, lime resin, beeswax, carnauba wax, and natural driers. Protects against water, dirt, and dust. Antistatic with extra surface hardening qualities.

MANUFACTURER'S STATEMENT
See INSTALLATION MATERIALS, p. 118.

MANUFACTURER

A.F.M. Enterprises Inc.
(American Formulating and Manufacturing)
1140 Stacy Court
Riverside, CA 92507
Tel (714) 781-6860, 781-6861   Fax (714) 781-6892

RELATED PRODUCTS

SUPER CLEAN, p. 200

DESCRIPTION

ONE STEP SEAL AND SHINE: Water-based modified acrylic latex polymer for use as floor sealer and polish. One-step system finish for hard surfaces (residential and industrial). Clear finish.

Resistant to detergent and chemical spills. May be buffed for high gloss finish. Recoating may be done without stripping the previous coat.

ALL-PURPOSE POLISH AND WAX: Water-based emulsion cleaning and polishing compound formulated without the use of toxic solvents. Contains paraffin wax and preservatives.

For use on furniture, vinyl floor coverings, all vinyl and metal finishes on automobiles and appliances.

Available in 2 types: liquid and creamy paste.

MANUFACTURER'S STATEMENT

See p. 152.

PRODUCTION

LOW EMISSIONS PLANT

PACKING / SHIPPING

MINIMUM, RECYCLED, RECYCLABLE PACKAGING

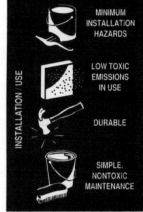

INSTALLATION / USE

MINIMUM INSTALLATION HAZARDS

LOW TOXIC EMISSIONS IN USE

DURABLE

SIMPLE, NONTOXIC MAINTENANCE

RESOURCE RECOVERY

OTHER

RESEARCH & EDUCATION PROGRAMS

PRODUCTION

SUSTAINABLY
ACQUIRED, OR
RENEWABLE
RESOURCE

LOW
EMISSIONS
PLANT

PACKING SHIPPING

INSTALLATION / USE

MINIMUM
INSTALLATION
HAZARDS

LOW TOXIC
EMISSIONS
IN USE

RESOURCE RECOVERY

OTHER

PRODUCT
All-natural thinner

MANUFACTURER
Biofa Naturfarben GmbH (Germany)
Imported by Bau, Inc.
P.O. Box 190
Alton, NH 03839
Tel (603) 364-2400

RELATED PRODUCTS
Biofa priming oils, p. 184
Biofa varnishes, p. 167
Biofa waxes, p. 188

DESCRIPTION
THINNER #0500: Pleasantly fragrant, natural thinner, free of synthetic solvents and aromatic substances. Suitable for thinning and cleaning natural resin and oil-based paints. For cleaning brushes, utensils, and removing paint spots. Oily dirt can also be removed. Do not use for water-based products.

MANUFACTURER'S STATEMENT
See p. 146.

PRODUCT
Thinner and solvent

MANUFACTURER
Livos Plant Chemistry (Germany)
Imported by
Eco-House (1988) Inc.
P.O. Box 220, Stn. A
Fredericton, NB E3B 4Y9
Tel (506) 366-3529
Fax (506) 366-3577

Livos Plant Chemistry / USA
1365 Rufina Cir.
Santa Fe, NM 97214
Tel (505) 438-3448

DESCRIPTION
#222 CITRUS THINNER: Thinner dissolves and thins all Livos oil base products. Clear, colorless liquid, with a mild scent of orange peel. Contains isoparaffinic hydrocarbons, a purified, odorless petroleum extract which is used as a solvent to dilute the citrus content. (Some people may be sensitive to high citrus concentrations.)

#710 KIROS ALCOHOL THINNER FOR SHELLACS: Pure ethyl alcohol distilled from beets and potatoes, and denatured with 3% pine oil. Designed for cleanup and thinning all Livos shellac products, as well as for conventional alcohol-based products. Mild pleasant fragrance, and less harsh than conventional methyl alcohol.

MANUFACTURER
See p. 147.

PRODUCTION

SUSTAINABLY
ACQUIRED, OR
RENEWABLE
RESOURCE

LOW
EMISSIONS
PLANT

PACKING / SHIPPING

MINIMUM,
RECYCLED,
RECYCLABLE
PACKAGING

INSTALLATION / USE

MINIMUM
INSTALLATION
HAZARDS

LOW TOXIC
EMISSIONS
IN USE

RESOURCE RECOVERY

OTHER

RESEARCH &
EDUCATION
PROGRAMS

PRODUCT
All-natural thinner for oil and resin finishes

MANUFACTURER
Naturhaus (Germany)
Imported by
Lindquist Marketing
P.O. Box 3542
33118 Whidden Ave.
Mission, BC V2V 2T2
Tel (604) 826-1547

Lindquist Marketing
115 Garfield St.
Sumas, WA 98295
Tel (604) 826-1547
Fax (206) 988-2207

DESCRIPTION
NATURAL THINNER #1: Unscented, colorless natural thinner, for use with oil and resin finishes, and some possible use with synthetic products. Use as a thinner for lacquers, linseed oil, varnish, and other natural resin oil products. Also used for cleaning surfaces, tools, and for removing excess wax. Contains orange turpentine and pine oil.

MANUFACTURER'S STATEMENT
See INSTALLATION MATERIALS, p. 118.

PRODUCTION

SUSTAINABLY ACQUIRED, OR RENEWABLE RESOURCE

LOW EMISSIONS PLANT

PACKING / SHIPPING

INSTALLATION / USE

MINIMUM INSTALLATION HAZARDS

LOW TOXIC EMISSIONS IN USE

RESOURCE RECOVERY

OTHER

FAIR BUSINESS PRACTICES

MANUFACTURER

Auro Organic Paints (Germany)
Imported by Sinan Co., Natural Building Materials
P.O. Box 857
Davis, CA 95617-0857
Tel (916) 753-3104

DESCRIPTION

NO. 191 PLANT THINNER: All-purpose thinner for natural oil-soluble products. Can be used to clean brushes or tools, and to remove spots of natural oil-soluble products.

NO. 219 PLANT ALCOHOL THINNER: All purpose thinner and cleaning solvent for alcohol soluble products.

NO. 461 PAINT STRIPPER: Alkaline finish remover paste without noxious solvents.

MANUFACTURER'S STATEMENT

See p. 145.

PRODUCTION

SUSTAINABLY ACQUIRED, OR RENEWABLE RESOURCE

LOW EMISSIONS PLANT

PACKING / SHIPPING

INSTALLATION / USE

MINIMUM INSTALLATION HAZARDS

LOW TOXIC EMISSIONS IN USE

RESOURCE RECOVERY

OTHER

RESEARCH & EDUCATION PROGRAMS

PRODUCTION

LOW
EMISSIONS
PLANT

PACKING / SHIPPING

MINIMUM,
RECYCLED,
RECYCLABLE
PACKAGING

INSTALLATION / USE

MINIMUM
INSTALLATION
HAZARDS

LOW TOXIC
EMISSIONS
IN USE

RESOURCE RECOVERY

OTHER

RESEARCH &
EDUCATION
PROGRAMS

PRODUCT
Low toxicity sealer/strippers for paint and films

MANUFACTURER
A.F.M. Enterprises Inc.
(American Formulating and Manufacturing)
1140 Stacy Court
Riverside, CA 92507
Tel (714) 781-6860, 781-6861   Fax (714) 781-6892

RELATED PRODUCTS
SUPER CLEAN, p. 200

DESCRIPTION
STRIPPER 66: Cold cleaning and film stripping, water-based, liquid solvent compound. Will remove 3 mil films of water-based type coatings and sealers.

Low toxicity water-based stripper does not contain methylene chloride, trichloroethane, perchloroethylene, or aromatic solvents. Contains a safe solvent necessary for stripping performance.

M P STRIPPER: Liquid tile stripper, comes ready to use, and requires no dilution. Can be rinsed with water. Noncorrosive, low flammability and non-hazardous. For use on porous or nonporous tile.

MANUFACTURER'S STATEMENT
See p. 152.

PRODUCT
Low toxicity stripper for paint and films

MANUFACTURER

3M USA
3M Center
St. Paul, MN 55144-1000
Tel (612) 733-1110
Fax (612) 733-9973

3M Canada Inc.
P.O. Box 5757
London, ON N6A 4T1
Tel (519) 451-2500
Fax (519) 452-6142

RELATED PRODUCTS

3M HEAVY DUTY STRIPPING PADS FOR FLAT SURFACES
3M HEAVY DUTY STRIPPING PADS FOR CURVED SURFACES
3M FINAL STRIPPING PADS FOR RESIDUE REMOVAL

DESCRIPTION

SAFEST STRIPPER PAINT AND VARNISH REMOVER: Contains water, dimethyl adipate, dimethyl glutarate, and hydrated magnesium aluminum silicate. Product is noncaustic, that is, no gloves are required, and won't harm wood. No special ventilation is required for use indoors or out, as no harmful fumes are emitted. The semipaste adheres well to vertical surfaces and remains active up to 30 hours. Nonflammable and cleans up with water.

MANUFACTURER'S STATEMENT

3M's corporate environmental policy, implemented in 1975, is to recognize and exercise its responsibility to solve its own environmental pollution and conservation problems, prevent pollution at the source wherever possible, develop products that will have a minimum effect on the environment, and conserve natural resources. 3M's facilities and products meet and exceed the regulations of all federal, state, and local environmental agencies.

Since the inception of the 3P program (Pollution Prevention Pays) in 1975, 3M's releases of air, water, sludge, and solid waste pollutants have been reduced by half, per unit of production. The program has also saved the company $500 million. The objective of the 3P program is to reduce all hazardous and nonhazardous wastes released into the air, land, and water by 90%, cutting the generation of all waste by 50% (from 1987 levels) by the year 2000. Many environmental achievement awards have been presented to the company for their efforts, among others, from the Environmental Protection Agency, World Environment Center, America's Corporate Conscience, and the National Wildlife Federation.

Energy efficient programs in effect are: carpooling/ferrying for employees, energy recycling, waste-to-energy schemes, and solar energy systems. Packaging design evaluation is ongoing to reduce the materials used, to research reusable, returnable materials, and use recycled materials wherever feasible.

3M has a policy of not marketing its products as "Green," as the company feels that environmental protection is a responsibility and not a marketing scheme.

PRODUCTION — IN-PLANT ENERGY EFFICIENCY & RECYCLING — LOW EMISSIONS PLANT

PACKING / SHIPPING

INSTALLATION / USE — MINIMUM INSTALLATION HAZARDS — LOW TOXIC EMISSIONS IN USE

RESOURCE RECOVERY

OTHER — FAIR BUSINESS PRACTICES — RESEARCH & EDUCATION PROGRAMS

PRODUCTION

SUSTAINABLY
ACQUIRED, OR
RENEWABLE
RESOURCE

LOW
EMISSIONS
PLANT

PACKING / SHIPPING

INSTALLATION / USE

MINIMUM
INSTALLATION
HAZARDS

LOW TOXIC
EMISSIONS
IN USE

RESOURCE RECOVERY

OTHER

RESEARCH &
EDUCATION
PROGRAMS

PRODUCT
All-natural cleaners and polishes

MANUFACTURER
Auro Organic Paints (Germany)
Imported by Sinan Co., Natural Building Materials
P.O. Box 857
Davis, CA 95617-0857
Tel (916) 753-3104

DESCRIPTION
NO. 411 PLANT SOAP: Pure potassium soap made of plant oils, with beeswax. Highly concentrated, mild, organic, for all surfaces, with slight regreasing effect.

NO. 415 CLEANING SOLUTION: Highly concentrated plant soap for heavily soiled surfaces. Can be used undiluted or diluted in warm water.

NO. 441 CEMBRA FURNITURE POLISH: For furniture care and cleansing. Prepared on a base of pure spirit, shellac and beeswax. Highland cembra oil provides insect repellent effect.

MANUFACTURER'S STATEMENT
See p. 145.

PRODUCT
All-natural soaps and cleaners

MANUFACTURER

Livos Plant Chemistry (Germany)
Imported by
Eco-House (1988) Inc.
P.O. Box 220, Stn. A
Fredericton, NB E3B 4Y9
Tel (506) 366-3529
Fax (506) 366-3577

Livos Plant Chemistry / USA
1365 Rufina Cir.
Santa Fe, NM 97214
Tel (505) 438-3448

RELATED PRODUCTS

#560 DRYADEN FURNITURE POLISH, p. 189
#565 ALIS FURNITURE POLISH, p. 189
#315 GLEIVO LIQUID BEESWAX, p. 189

DESCRIPTION

#551 LATIS ALL-PURPOSE SOAP (LIQUID): Organic soap for tiles, plumbing fixtures, counters, wood floors, and others.

#595 VELIO FURNITURE CARE SET: May be used on all finishes except shellacs. Cleans and improves scratch resistance. For use on wood, cork, linoleum, and terracotta surfaces. It reduces electrostatic loading and the attraction of dust.

#550 TEKNO WAX REMOVER: Diluted, cleans oiled and waxed wood, and cork surfaces. Undiluted, removes waxes and oils.

MANUFACTURER'S STATEMENT

See p. 147.

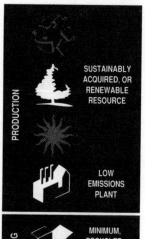

PRODUCTION

SUSTAINABLY ACQUIRED, OR RENEWABLE RESOURCE

LOW EMISSIONS PLANT

PACKING / SHIPPING

MINIMUM, RECYCLED, RECYCLABLE PACKAGING

INSTALLATION / USE

MINIMUM INSTALLATION HAZARDS

LOW TOXIC EMISSIONS IN USE

RESOURCE RECOVERY

OTHER

RESEARCH & EDUCATION PROGRAMS

LOW
EMISSIONS
PLANT

PRODUCTION

MINIMUM,
RECYCLED,
RECYCLABLE
PACKAGING

PACKING / SHIPPING

MINIMUM
INSTALLATION
HAZARDS

LOW TOXIC
EMISSIONS
IN USE

INSTALLATION / USE

RESOURCE RECOVERY

RESEARCH &
EDUCATION
PROGRAMS

OTHER

PRODUCT
Low toxicity all-purpose cleaners

MANUFACTURER
A.F.M. Enterprises Inc.
(American Formulating and Manufacturing)
1140 Stacy Court
Riverside, CA 92507
Tel (714) 781-6860, 781-6861   Fax (714) 781-6892

DESCRIPTION
SUPER CLEAN: Biodegradable all-purpose cleaner. Eliminates the need for several "specialty" cleaners. Noncorrosive, for use on floors, woodwork, walls, counter tops, tubs, tile, furniture, rugs, etc.

Noncaustic and nonirriting to the skin.

SAFETY CLEAN: Active cleaner with antibacterial properties. Effective against a broad spectrum of fungal and bacterial growth, for use in tub enclosures, shower stalls, water closets, saunas, etc.

Provides an antibacterial protective film after use.

CARPET SHAMPOO: Low foaming, water-based carpet shampoo. May be used as a general household cleaner.

Deodorizes and has long-lasting germicidal properties.

MANUFACTURER'S STATEMENT
See p. 152.

FURNITURE
AND ACCESSORIES

Furniture is fundamentally different in many respects from the other categories of materials in this guide. And because it is so complex, discussing it brings together many of the issues raised in other sections. Furthermore, furniture, unlike carpets, paints, and other interior materials, is in part produced by a vast number of small manufacturers and sold locally. Only the major furniture manufacturers sell their products nationally and internationally.

Because of this complexity, it is difficult to report on furniture in the same manner as on some of the simpler materials. In your area you will be able to find only a few of the many companies described in the following pages. Most of the residential furniture listed uses teak or other woods from managed plantations. The rest consists of office furniture systems. Because furniture is bulky and heavy to ship there are important advantages to shopping locally, above and beyond the value of your support for local businesses. For these reasons this introduction is more detailed than any of the introductions to other materials.

PRODUCTION AND MANUFACTURE

Most furniture companies are secondary manufacturers to a greater degree than many other industries because they buy manufactured materials and assemble them. Only a few, such as all-wood manufacturers, and cane and rattan makers do most of their own material processing. The environmental impact of manufacture is therefore divided among many other plants which supply the furniture maker. Some of the important steps in the manufacturing process are:
- Gluing and laminating
- Welding and metal plating
- Machining
- Painting, staining, and waxing
- Polishing
- Packaging and shipping
  Each of the materials purchased by the manufacturer also has its own envi-

ronmental costs. Some of these are connected to:

- Logging
- Wood products manufacture (plywood, and other composite wood boards and fiberboards)
- Smelting and refining of metals, glass, and ceramics
- Plastics and plastic foams production
- Fiber and textile production and dyeing
- Leather tanning
- Paint and adhesive production

Each of these is discussed below under MATERIALS, UPHOLSTERY TEXTILES, and PAINTS AND FINISHES (pp. 203–18).

ENVIRONMENTAL AND HEALTH IMPACTS OF MANUFACTURE

The major impacts incurred during manufacture of the actual furniture are air and water pollution, toxic waste, potential physical and chemical injury to workers, and energy and resource consumption. Of these, perhaps the most serious is the environmental impact of finishing procedures.

Fifty years ago most furniture was made from woods and upholstered coverings or leather, with some metal and glass parts. Fillings were often spring frames with cotton, jute, or other natural fiber padding. Finishes were generally linseed oil or varnish bases which were applied by hand. Fine furnishings were lacquered, and some metal parts painted, plated, or enameled. These were not all environmentally safe, or healthy practices, however, because many traditional finishes contained toxic metals and solvents, and tanning and metal-plating operations were serious polluters.

Today most furnishings are made from wood veneers or plastic laminates on particle board cores, vinyl or polyester fabric covers, and polyurethane foam fillings. Most finishes are synthetic resin-based and are applied by spraying. Though toxic metals have been restricted in finishes by law, the chemical makeup of all furniture parts is now vastly complicated by technical innovation.

SOLVENT RECOVERY AND WATER-BASED FINISHES

Coming in contact with solvent-based finishes has always been the major health risk to workers in this industry. The solvents are also the main source of toxic waste. With the use of advanced vapor collection methods and solvent recovery systems, a plant can be a safer place to work and can produce less toxic waste. Many manufacturers have found that these systems have a rapid payback period, because solvent recovery programs substantially reduce solvent purchases and toxic waste.

Advances in water dispersion of finishes have also led to high performance materials with far less solvent content, and with minimal quantities of waste from cleaning equipment. Even the automobile industry, where finishes must be very durable, is beginning to switch to water-based finishes. Water-based finishes are usually safer to handle.

POWDER-COATED METALS

Furniture can also contain many plated metal parts. Platings are usually made of cadmium, zinc, or chromium. Though plating may be done for esthetic reasons, many parts such as drawer slides and shelving frames are plated for corrosion resistance. Both cadmium and chromium plating are hazardous industries which create serious toxic waste problems. A major advance in protective metal coatings is the powder-coating process. In this process metals receive a thin coating of dry powdered polymer which is "sprayed" using a static electrical charge to make it cling to the parts. It is then heated and fused to make a durable and corrosion-resistant finish. No solvent is used, no hazardous waste is produced, and worker exposure is minimized. (See also POWDER COATINGS, p. 218.)

OZONE-SAFE UPHOLSTERY FOAMS

Another important innovation in furniture manufacture is the introduction of ozone-safe upholstery foams. Until 1990 most high performance upholstery foams were manufactured with ozone depleting chlorofluorocarbons (CFCs) or with toxic methylene chloride. These chemicals were used as a "blowing agent" to expand the foam's closed cells, giving it structure and resilience. Though these gases perform well, they are eventually lost to the atmosphere. Here they are significant polluters, contributing to ozone layer depletion, which exposes the earth to excess ultraviolet radiation.

Under international agreement, all CFCs used for this purpose must cease in the next five years. Replacement blowing agents are now in use, such as carbon dioxide and nitrogen, which are both ozone friendly and safe for those who handle them. Some manufacturers provide a label on padded items which indicates that the foams used are ozone safe.

MATERIALS

SOLID WOODS

Some traditional and handcrafted furniture is still made from solid woods, usually hardwoods (oak, maple, beech, ash, mahogany, and teak). The main use of solid softwoods (pine, fir, and spruce) is in American Colonial reproductions, Scandinavian pine furniture, and futon bed frames.

FOREST MANAGEMENT

The major environmental concern with hardwoods is the source of the woods. It is now well known that much of North America's imported tropical hardwood is obtained by highly destructive land clearing in sensitive tropical zones. These practices are driven by global market conditions and economic desperation in many of the countries near the equator.

Poor forest management is, however, a serious concern in nearly every region. In the U.S. and Canada destructive practices are causing severe soil erosion, loss of wildlife habitat, damage to rivers and lakes, and the loss of natural wilderness value for future generations. In the equatorial rainforest regions, environmental damage also means the displacement of native peoples, and, due to fragile tropical ecosystems, the loss of fragile soils and thousands of rare species of plants and animals.

Deforestation also leads to changes in weather patterns, the expansion of deserts, and further global warming, because the forest plays such a large role in absorbing moisture, regulating sunlight and wind, and absorbing carbon dioxide from the atmosphere. Be cautious of claims about "sustainable woods." There are no widely accepted standards for sustainable forestry practice.

TYPES OF WOOD

Most hardwoods are slow growing and cannot sustain rapid increases in demand. Softwoods mature much faster and can be managed more sustainably. For example, replanted pine in the southern U.S. can be cut in only 25 years.

SOFTWOODS: cedar, cypress, fir, hemlock, larch, pine, redwood, spruce.

Generally, only a few of these woods are used for interior furnishings. These are typically pine or fir pieces, in Swedish and New England styles. Most of these species, however, are used for building construction. Red cedar and redwood are the slowest to regenerate and the remaining forests are rapidly diminishing. They have the advantage of natural rot resistance and are popular for patio furniture. Very durable outdoor furniture is arguably an appropriate use of these naturally rot-resistant woods, particularly if it is well cared for.

DOMESTIC HARDWOODS: alder, apple, ash, aspen, beech, birch, cherry, elm, gum (eucalyptus), hickory, lime, maple, oak (red and white), pear, poplar, sycamore, walnut, whitewood.

Most domestic hardwoods are used in traditional residential furniture, cabinet manufacture, and on edges and trims of pieces containing mainly veneers. A few are used as flooring (see FLOORING, p. 79). Beech, cherry, and walnut are the most depleted of the domestic hardwoods, although there are isolated sustainably managed forests in New England and the southern U.S.

One of the most important environmental advantages of using domestic hardwoods is simply the practical position that if we use something from our own region, we have direct appreciation and experience of its value.

TROPICAL HARDWOODS: Seriously depleted tropical hardwoods include aformosia, ebony, gaboon, iroko, jelutong, kapur, kempas, keruing, lauan (also called Philippine mahogany, used for hardwood plywood), mahogany, merbau, padauk, ramin, rosewood, sapele, teak, tulipwood, utile.

Alternatives which are better managed through replanting and harvesting controls are greenheart from Guyana; virola from the Amazon; teak from plantations in Java, Thailand, and Burma; and rubberwood from Malaysia.

Some manufacturers have stopped buying threatened hardwoods and are switching to better managed alternatives. One U.S. furniture manufacturer in particular is experimenting with several of the hundreds of tropical hardwoods which are now wasted in the process of logging species that are in demand on the world market. If these woods become valued it will encourage better forest management.

Some veneer manufacturers are producing interesting colored and textured veneers from low grade woods that have been wasted in the past. Domestic woods, such as alder from the Pacific northwestern U.S. and western Canada, and poplar and birch from the midwestern U.S. and Canadian prairie, are currently little utilized, but offer good potential for custom furniture, cabinets, and interior finishing.

TROPICAL HARDWOOD LABELING PROGRAMS

It is very difficult to ascertain if an imported tropical hardwood has come from a managed forest. Some will argue that there are no sustainably managed forests. You can, however, patronize manufacturers who do not use threatened species and have tried to verify that their hardwoods come from managed forests. The best assurance that an imported wood has come from a plantation where conservation and replanting are practiced, and where workers are likely to be treated fairly, is the presence of either the "Smart Wood" or "Good Wood" label. These are administered by the Rainforest Action Network and Friends of the Earth, respectively, who provide independent research and monitor the programs.

WOODS AND HEALTH

Softwoods contain volatile resins called terpenes. Terpenes are part of the nutrient and circulatory system of evergreen trees, which must survive frosts while retaining their foliage. All woods from evergreens contain some degree of oily, sticky, or hard residue from terpenes which give the wood its characteristic sharp odor. Some people are quite allergic to terpenes and will be immediately affected by the presence of softwoods. The fresher the wood and the more exposed its ends and edges, the more terpenes are released. For some people, only hardwoods, or well-aged or sealed softwoods are acceptable indoors. Some hardwoods also contain resins which, though they may not have a strong odor, are very toxic to inhale. For this reason, woodworkers should be vigilant about using respirators and having adequate dust collection equipment when sanding and cutting woods.

TWIG OR STICK FURNITURE

The Appalachian bent twig furniture style has made a comeback recently, and several small manufacturers and mail order services now offer this furniture. It is made from green branches which are bent or woven into elegant pieces. The bark is often left on. This is clearly an efficient use of resources, as it does not require logging or sawmilling, and can make use of waste wood from thinning managed forests and from pruning.

WICKER

True wicker is made from twigs of willows which grow in northern temperate climates, and wicker work has been an established craft in Europe for millenia. Like bent twig or stick furniture, it is a sustainable use of resources because wicker harvesting does not destroy the tree.

RATTAN AND BAMBOO

Rattan furniture is made from several hundred species of palms which grow only under the canopy of a tropical forest. If the forest is cut they are lost, and the palms have already become a depleted resource in many areas due to deforestation. Rattan is also in short supply in some markets due to an export ban on the raw material imposed by Indonesia in 1988. The purpose of the ban was to encourage manufacturing of furniture in Indonesia, and it has been successful in the past few years in developing a better quality Indonesian product, while creating a value-added export product for that country. Until the export ban, most of the high quality furniture was manufactured in the Philippines from Indonesian rattan. Rattan harvesting is an important sustainable use of tropical forests as it does not require logging or roadbuilding to the same degree as wood production.

Bamboo, a fast growing giant grass, is in plentiful supply in most tropical and subtropical areas and is not likely to be depleted. Sturdy, inexpensive furniture is manufactured from bamboo and split bamboo in China, southeast Asia, and India.

COMPOSITE WOOD BOARDS

The majority of inexpensive furnishings, and most of the lines of commercial wood veneer or plastic laminate furnishings are built with frames and cores of composite wood boards. Often called "manufactured wood products," these materials are made from wood dust or chips which have been pressed into sheets using glues to give them strength and surface hardness. These materials use wood resources very efficiently, and provide a stable and easily machined base for veneers, paint or lacquer finishes, and plastic laminate surfaces. They are, however, generally made with urea-formaldehyde and phenol-formaldehyde based glues, which make handling and machining a toxic exposure risk for people who work in the industry. Furthermore, if the formaldehyde-based materials are not chemically stabilized, or completely sealed by veneer or special sealers, they will offgas formaldehyde and other irritating gases which add to indoor air quality problems.

In fact, it is now very common to find complaints in homes and offices of burning eyes, congested sinuses, and itchy skin which are related to recently installed new furniture made from low quality manufactured wood products which have not been sealed.

PARTICLE BOARD

Particle board is a general term for manufactured wood products which are made from softwood dust and relatively large quantities of glue pressed into panels. They are classed as medium density fiberboards (MDF). Some of the most common types are made from pine, hemlock, or red cedar dust with urea-formaldehyde glue. The glue may comprise over 15% of the material by weight. These boards use waste wood material from sawmills which would otherwise be burned or landfilled. There is an obvious environmental merit for this type of resource use.

This product is widely used for kitchen and bathroom cabinet sides, backs and doors, countertop and desktop substrates, shelving cores, and flooring underlayment. When used as a substrate where the surfaces will be thoroughly sealed by an impermeable covering material, such as plastic laminate or veneer, the formaldehyde gas emissions from the material are dramatically reduced. However, unsealed edges, bottoms, backs, and sides, and even holes drilled for hardware which have not been plugged will allow formaldehyde release.

Several types of alternative materials are available. Some particle boards are made with exterior glues which emit less formaldehyde (classed as "exterior" or "exposure one" materials); others are chemically stabilized to meet more stringent emissions standards (classed as "low emission" or "European standard" materials); and some are made using no formaldehyde-based glues at all (classed as "formaldehyde free" materials).

Health Note: The dusts from working with particle boards can be very irritating. Some people who handle them regularly will become sensitized and may develop allergic reactions. Whenever particle board components are used for interiors it is best to have all cutting and machining done in the fabricator's shop where the dusts can be carefully controlled.

In most limited budget interiors it is very difficult to avoid the use of particle board core cabinets and furnishings. When faced with this dilemma, it is still possible to select products which are likely to cause the least air pollution.

Check first with manufacturers to find out if they have done formaldehyde emissions testing on their materials. If not, look for manufacturers who use "formaldehyde free" or "European standard" board, or "exposure one" board rated under U.S. emissions rules. Check also that all particle board surfaces are completely sealed with veneers, especially edges, backs, under desk tops, and inside drawers. If deficiencies exist in the units you would like to use, consider requesting that a liquid sealer (formulated for formaldehyde containment) be applied to exposed particle board surfaces, or apply it yourself. Formaldehyde-free boards and sealers are listed in INTERIOR CONSTRUCTION PANELS and FINISHES, pp. 43, 158, 168, 173, 174.

HIGH DENSITY HARDBOARD (WOOD FIBERBOARD)

Hardboard or fiberboard are common terms for manufactured wood products made from ground wood pressed into a thin, very strong panel using high temperatures. These products can be made without glues, due to the natural ability of the lignin in wood to bind the fibers together. However, they often contain glues or additives to give them special properties, such as in the surface treated or "tempered" varieties which have a very hard and glossy surface.

Commonly used for furniture backs, drawer bottoms, shelf surfaces, and door facings, this material is also flexible and is often sold prepunched for pegboard installations. (See INTERIOR CONSTRUCTION PANELS, p. 35, for more information.)

ADHESIVES AND FASTENERS

Common white glues such as polyvinyl acetate (PVA) and modified PVAs are adequate for assembly of all wood composition materials for indoor use. These are the safest and lowest in environmental cost of the common glues. Dowels, screws, and nails (used in conjunction with glues), and clamps and special fasteners provide a sturdy end product with a minimal use of glue.

LAMINATING

Particle board (MDF) is used and favored as a substrate for laminates and veneers because of its smooth surface and ease of workability. Low pressure laminates (melamine, high density phenolic and polyesters) are either heat bonded or contact cemented to the surface. (See PLASTIC LAMINATES below.) Solvent-based contact cements, water-based polyvinyl or casein-latex glues can be used for laminating. The solvent variety is quite flammable and toxic to handle, yet generally performs no better than the safer, water-based variety. (See INSTALLATION MATERIALS, p. 105, for more information.)

FORMALDEHYDE REDUCTION METHODS

To reduce emissions, several American and European particle board manufacturers are using an anhydrous ammonia treatment to neutralize the formaldehyde content of boards. Ammonia fumigation is also sometimes used for interiors with high formaldehyde emissions (such as mobile homes), and can be done right inside the structure. The ammonia used is a strong toxic solution (not the household variety), and must be applied by professionals. The process results in a fine dust of hexamethylene tetramine (methenamine), which must be thoroughly cleaned away before the boards are used or the space is reoccupied. Substantial amounts of this compound will remain in the wood product itself. It is harmless if it is not released through cutting or machining.

Formaldehyde emissions begin to decrease as soon as the boards are produced, decreasing continually with time, so that older materials produce fewer emissions than newer ones. Coatings and laminates can reduce emissions by up to 95%, but all surfaces must be sealed for this to be effective. The emissions which are "contained" by this method are released more slowly, extending the decay period.

Particle board manufacturers may also add formaldehyde "scavengers," such as bisulfite compounds, during the manufacturing process. Scavengers combine with the free formaldehyde to form stable chemical compounds that will not leave the board.

VENEERS AND PLYWOODS

Made from logs which can be "peeled" into thin sheets (called veneer) and then glued onto a substrate, plywood is sometimes used for cabinet sides and backs, and in chair and sofa construction. (See INTERIOR CONSTRUCTION PANELS, p. 33, for more information.)

The use of fast growing and previously less utilized domestic hardwoods, such as birch and alder, as both the corewood and the face veneer is an important step in reducing consumption of threatened tropical forests. Another important step in the management of tropical woods is to begin using little known woods which are now being wasted in the process of logging the highly prized teak, mahogany, and rosewood trees. Some furniture manufacturers are experimenting with woods which currently are said to have no market value in the tropics. This is an important step for two reasons. First, it encourages conservation by increasing the value of the forest. Second, it provides more potential for a diversified income base for the people of the tropical regions. Look for the "Smart Wood" or "Good Wood" labels on imported hardwood products, and the introduction of little known species.

An important and more sustainable alternative to high grade natural veneer is "reconstituted" wood veneer. Reconstituted veneers are made from lower valued woods, such as poplar and aspen, and are machined and textured, and sometimes dyed to produce interesting effects. One method uses laminated layers which are recut (sometimes several times), to create a consistent and fully usable finish material. Reconstituted veneers conserve valuable hardwoods by utilizing lower valued woods and by using more of the material in the process.

While all plywoods have emission patterns similar to particle boards due to their glue content, the quantity of glue is generally less, and the type of glue used in exterior grade plywoods is lower in emissions than that used in standard interior particle boards. Interior plywoods, such as veneered wall paneling, often use interior grade glues which are formaldehyde sources, though again less glue is present than in particle boards. Exterior grade plywoods are the safest from an emissions standpoint, and emissions can be further reduced by sealing surfaces with veneers, or varnish and lacquer coatings. (See FINISHES, p. 143, 158, 168, 173, 174.)

HIGH PRESSURE LAMINATES (HPL)

HPL, also called phenolic laminates, are the most popular surfaces for inexpensive furniture and cabinets. A sheet material is formed from multiple layers of kraft paper saturated with phenolic resin, a decorative layer containing colors and patterns, and a very thin top sheet of paper saturated with melamine resin. These are fused together under high temperature and pressure to produce a stiff, durable sheet. The final product is relatively chemically stable, and can seal in formaldehyde emissions from particle board and plywood, the most common substrates used with laminates. Laminate manufacture involves producing and handling toxic materials and toxic waste; the dusts from machine cutting are hazardous; and the glues used to install it are hazardous. It is, however, an appropriate use of plastics, in the final analysis, because a very small amount of plastic is used. The product is very long lasting, and it effectively serves the need for an inexpensive, durable, washable, and sanitary surface.

LOW PRESSURE LAMINATES (LPL)

A preprinted or solid color decorative paper is saturated with a resin such as melamine and fused, under heat and pressure, to a substrate such as particle board or plywood. No additional adhesive is required. Low pressure laminates are common as inexpensive wall paneling, door panels, shelving, and cabinet faces, but they are not as durable as high pressure types, nor as chemically stable. Gases from the substrate are not sealed in as well and are able to escape into the air.

MELAMINE LAMINATE

In this process, a melamine resin system is used to saturate the paper laminate and bond it to the substrate. The common resin used is melamine formaldehyde, a thermosetting urea resin with exterior capabilities. It is chemically stable and has low emissions, but it produces hazardous dust when being machined.

POLYESTER SOLID SURFACING

This material is a thick, solid matrix of polyester resins and dyes with mineral fillers. Several variations of this material are available using different polymers and fillers. It provides a very stable, durable, and sanitary surfacing material for higher priced cabinet tops and furniture. The production process uses large amounts of resin in comparison to high pressure laminates, but once installed and in use it has minimal indoor air pollution potential.

ACRYLIC SOLID SURFACING

This material is a thick, solid matrix of acrylic resin containing dyes, marble chips, and other decorative fillers. It provides a very stable, durable, and sanitary surfacing material for higher priced cabinet tops and furniture. It is, however, less scratch resistant than other composition materials. Like polyester surfacing materials, the production process uses large amounts of resin in comparison to high pressure laminates, but once installed and in use it has minimal indoor air pollution potential.

SOLID PLASTIC FURNITURE

A lot of playroom and garden furniture is now available in solid plastic. Most is made from molded polyvinyl chloride (PVC), which has serious environmental and health risks from its manufacture. The raw material for the plastic-making process, vinyl chloride, is highly carcinogenic, though industry workers are rarely exposed because it is a "sealed process." Plastic furniture is not durable or repairable, and though pure PVC is recyclable, there are few facilities for handling it.

METALS

Metals are used for furniture frames, upholstery springs, and table bases. Steel is the most common material for this use, though aluminum is found in lightweight chairs and furniture. Steel, if used in an exposed location, is usually either plated, vinyl-clad, or painted. Aluminum is often left bare. Other metals, such as copper and brass, are used for decorative purposes.

The manufacture of steel and copper has several negative environmental impacts. For steelmaking, coal is first "coked" to reduce it to carbon, then fed into a furnace containing iron ore and limestone. It is further refined through the injection of oxygen. The result, steel, is then formed into shapes, or alloyed to make other metals like stainless steel. The main environmental concerns of steelmaking are the waste products from coking, and the air and water pollution, and solid waste from mills. Though in the past twenty years the steel industry has made major advances in controlling pollution, it is still a high impact industry. Stainless steel which is alloyed with nickel, chromium, and other metals has a much higher environmental cost than other steel, due to the high energy use and emissions from the refining of the other metals. Steel which has chromium or other protective and decorative plating is put through a plating process, which is a heavy energy consumer and producer of hazardous waste and water pollution. The acids and toxic metal residues from plating plants are a chronic disposal problem which have been difficult to manage.

Copper smelting, for the production of copper and bronze, is also a highly polluting industry that releases acids and sulfur dioxide into the environment. One of the major problems with copper is that it is found in ores at such low concentrations that a great deal of waste is produced by refining.

Aluminum smelting is also a high impact industry. The process consumes large amounts of energy, producing waste problems, such as petroleum tars (from making carbon electrodes), and fluoride emissions.

Though metals are typically only used in small amounts in residential furnishings, they are more common in office furniture. As they have significant environmental impacts, it is fortunate that they are highly durable and, if not alloyed with other metals, highly recyclable. Alloyed metals are more difficult to recycle because they must be separated by material type. Plated metals are not as difficult to recycle, because the quantity of plating is small, but toxic metals, such as cadmium and nickel, can be released in the process. The best choices for metal furniture parts are bare aluminum, and steel which is painted or powder coated (see p. 218). All metals are safe from the standpoint of indoor air quality and emissions. Paints used on metals are usually cured at high temperatures, making them low emission finishes.

Glass is a significant element of furnishings. It is made in a high temperature process from silica sand, lime, and various fluxes, which are fused in a furnace and then cooled. The process demands a great deal of energy, primarily natural gas, and produces some air and water pollution. There are vast supplies of the raw materials needed for glass manufacture.

Though glass is technically recyclable, only recycling facilities for container glass are widely available, and the environmental benefit is marginal. It takes nearly as much energy and produces as much (or more) pollution to make glass from recycled material as it does to make it from new material. The best reason for recycling glass is to help solve the disposal problem. Some manufacturers are now making glass products, such as glass blocks and floor tiles, with waste glass. This demonstrates a case where converting the waste into another material, instead of back into its original form, has more environmental benefit than strict recycling. Sheet glass does not contain recycled materials because it has much more critical formulation requirements and cannot be allowed to contain certain impurities.

Glass is a highly chemically stable material, and will not deteriorate significantly with age or produce any emissions indoors. Though it is made with a process which consumes a great deal of energy and produces some pollution, it is a good choice from an environmental standpoint due to its durability and the availability of raw materials.

UPHOLSTERY TEXTILES

Textiles are another important component of furniture. Upholstery textiles may be made from synthetic fiber such as nylon and polyester, natural fiber such as wool, silk, and cotton, or blends of two or three types of fiber.

NATURAL FIBERS

WOOL

Wool is sometimes used in residential upholstery fabrics, though it is more widely utilized in contract or commercial fabrics. Raising sheep for wool is generally a benign activity if the grazing land can support them, and the energy requirements, and impacts on soil and water are much lower than those involved in growing cotton. Some environmental concerns and ethical questions regarding the treatment of animals are the dipping of sheep in pesticides to control parasites, the use of chemical treatments to remove wool from the skin of slaughtered animals, and the wool refining and dyeing processes. Wool is a renewable resource, but limited by the ability of the land to support sheep.

Wool fiber is very durable and naturally fire resistant, which reduces the treatments required to make it meet performance standards and fire codes. It is, however, prone to attack by moths, and treatment is required by law for export products. Mothproofing is achieved with toxic substances such as sodium fluorosilicate, which can cause skin irritation to some people. The fiber is naturally abrasive and contains natural oils (lanolin). Skin sensitivities to wool are quite common due to these properties. (See CARPETING, p. 50, for a further discussion of wool.)

Silk is made from the fibers of cocoons produced by cultivated silkworms maintained on a diet of mulberry tree leaves. It is a very labor-intensive process done almost exclusively in China, Thailand, and a few other areas in southeast Asia. Silk production has quite a low pollution impact until the stage of dyeing the thread or the finished product because it is essentially a low technology agricultural industry with little mechanization. There are many poorly controlled dyeing processes in producer regions, which release toxic metals, acids, and other hazardous materials into the water and soil. Some of the dyestuffs used may also be health risks to both the people who handle them and the end user. Because there is very little reliable information on the diverse silk producers, only products which are not dyed or are very light in color can be reasonably assumed to be free of toxic dyes.

PLANT FIBERS

Plant fibers used for fabrics and fillings are cotton, linen, ramie, kapok, and jute. Cotton is the most intensively farmed of all plant fibers. It depends heavily on the use of chemical pesticides and fertilizers for high yields. Cotton is vulnerable to attack from a large variety of pests, and it makes heavy demands on soil nutrients. Chemical defoliants are often used at the harvest stage to make the crop easier to pick.

Domestic cotton production is not severely regulated for the toxicity of the agents used because it is classed as a fiber crop, and the seeds are generally used as animal feed. In poorer countries, where cotton may be an important cash export item, there may be even fewer controls over pesticides and fertilizers used. DDT and other hazardous organochlorine groups, banned in industrialized countries because they accumulate and are toxic to birds and fish, are still in use in many countries, exposing workers and wildlife to serious health risks.

The production of cotton fiber also requires substantial amounts of energy for farm equipment and processing plants, but still uses less energy than that required to produce the same amount of synthetic fiber.

Linen is made from the spun fiber of the flax plant. Flax cultivation also uses fertilizers and pesticides, though to a lesser degree than cotton. Linen is the only natural fiber other than cotton which is widely cultivated by mechanized farming methods for textile use. Most other fibers are coproducts from crops, either grown in a semicultivated state, or collected in the wild.

Ramie is the fiber from a nettle plant which grows extensively in south Asia. Kapok is a cotton-like fiber from a tree which is found in southeast Asia. Jute is a coarse, tough fiber from an Asian shrub. Bangladesh is the largest producer country worldwide. There is very little information available on the cultivating and processing of these three plant fibers, though those grown on plantations may be subjected to pesticide treatments. All natural fibers may be allergenic to a few sensitive people, though allergies to cotton and linen are very rare.

SYNTHETIC FIBERS

There are two basic types of synthetic fabrics: the cellulosics (those made from plant fibers and wood), and the noncellulosics (those made from petroleum and glass).

These are fibers spun from resins derived from plant and wood cellulose. Cellulose is a carbohydrate, which is the structural material of all plants. It is called cellulose because it forms the walls of plant cells. When it has been digested by a chemical process similar to turning pulp into paper, it can be turned into plastic materials (viscose fibers), including rayon, viscose, and acetate. The major raw material sources (mainly eucalyptus and spruce) are renewable and fast growing. However, mills which break down logs into fiber are major sources of air and water pollution. Poor forest management practices, such as clearcut logging, are still the norm in the pulp and paper industries. There is a resource-use advantage to making cellulosics: they can be made from agricultural waste, such as cotton lint, instead of virgin wood fiber.

In the process of making viscose, the separated cellulose fibers are digested by carbon disulfide and sulfuric acid. These are hazardous materials in themselves and produce sulfate effluent, which is one of the major pollutants from pulp mills.

Cellulosics are sometimes tolerated better than noncellulosics or the more allergenic natural fibers, such as wool, by people with allergies or very sensitive skin. This is an important consideration in selecting furniture coverings which will be in contact with skin.

NONCELLULOSICS

These are fibers spun from resins derived mainly from petroleum, though some are made from glass. The main noncellulosics are polyester, nylon, acrylic, polypropylene, and fiberglass. Most of the raw material is petroleum which has been distilled into the basic molecular "building blocks" of the synthetic resins, and then linked together, by catalytic processes, into resins. These manufacturing processes occur mainly inside sealed equipment, but the waste materials, air pollution, and risks to workers handling hazardous materials are all serious environmental and health problems in the industry.

Fibers spun from glass and graphite do not involve the same degree of industrial hazard and pollution problems, because they are simply made from heated sand and lime, but the fibers have very limited uses in textiles. Fiber released into the air, however, is a serious respiratory hazard. Though glass fiber fabrics are used for some fire-resistant draperies, acoustic panels, and wall coverings, the majority of glass fibers are used as insulation materials and as reinforcement for flexible plastics (fiberglass).

FABRIC TREATMENT AND MAINTENANCE

Fabric treatment and maintenance is another environmental cost and health risk associated with furniture. Porous fabric coverings will readily trap dust and dirt from skin and soiled clothing, food spills, animal dander, and stains. Depending on the type of fabric and whether it is removable, the appropriate cleaning methods will be hand washing, machine washing, steam cleaning, or dry cleaning. Hand or machine washing can be done with soaps, washing soda, and other benign cleaners. Steam cleaning usually requires solutions which are somewhat less benign, residues of which remain on the fabric and can cause skin irritation. Dry cleaning uses such toxic solvents as perchloroethylene which deplete atmospheric ozone, are health hazards, and become hazardous waste when no longer useful.

Porous textile coverings are often treated with stain-resistant agents to reduce maintenance. These are generally paraffin-like resins dissolved in a solvent such as 1,1,1 trichloroethane, and they are applied by spraying, or during machine cleaning. The resin penetrates the fiber and sets, making it less porous and therefore less able to absorb stains. Though these agents can reduce cleaning requirements, they are also environmental risks and toxic to handle. The residues left on fabric can irritate those with sensitive skin.

Fabrics which are intended to retain a crease or to be more wrinkle resistant are also treated with either temporary or permanent finishes which change the structure of the material. One of the most common commercial finishes is a formaldehyde and latex resin which will stay in the material through many washings. Such finishes may be irritating to those with sensitive skin. Recent innovations in fiber blending have produced synthetic blends which will retain their shape after heating without chemical treatments. Mercerizing of cotton is another treatment used to permanently increase fabric strength and surface gloss. The process uses ammonia or caustic soda, both of which are health risks for workers and produce hazardous waste, but the result is a permanently treated fabric with no irritating properties.

Sizing is a temporary resin treatment to make the fabric hold its shape while being cut and stitched. It is usually a starch or water-soluble resin which can be removed with repeated washings, especially with washing soda (sodium carbonate).

WORLD TEXTILE MARKETS

North America has dramatically increased imports of fabrics from poorer textile-producing countries that have been able to profit from low cost and refugee labor. In many of these countries air pollution and effluent controls are not as strict as in North America and Europe, which often keeps production costs lower. Domestic jobs in farming and manufacturing are lost each day to this global market shift.

Third world countries need the economic infusion that the North American market can provide, but this infusion must not sacrifice or overlook social and environmental factors. In the past, fiber production has increased in these countries in the face of falling prices, creating further dependance on imported chemicals and contributing to the degradation of the land. Plant fibers are often planted as a cash crop in areas where people are starving and cannot find land to grow food for themselves. Consider asking your textile supplier about the source of raw materials and the trade practices of the industry.

LEATHER

Most of the leather used for upholstery comes from animals raised and slaughtered for meat. Although it prevents the waste of that part of the carcass, some people wish to avoid using animal products altogether.

The tanning and dyeing processes for making finished leather from hides have serious environmental consequences and health risks due to alkalis, acids, chromium compounds, and other toxic waste discharges. Tannery discharges are now fairly well regulated in most countries, but serious abuses exist in India, Morocco, Brazil, and other poorer nations. In India, unregulated tanneries provide a cheap service to overseas companies which ship raw hides in and finished hides out, while local environments and the health of workers suffer.

Furniture suppliers should be aware of the source of leather used in their products. Leathers which are processed in a region with substantial environmental regulation of tanneries should be a minimum requirement of an environmentally aware choice. The industry as a whole will eventually be alerted to the importance of these concerns when sufficient attention is paid to them by the consumer.

Highly finished leathers generally have dyed and polished surfaces which introduce more environmental and health risks in their production than unfinished leathers. However, maintenance becomes an easier task. Suedes and chamois, which absorb oils and soil, may require chemical cleaning, adding to the environmental burden. For most furniture uses, a leather with a low maintenance finish will have the least environmental cost over its lifetime.

A final important concern is the trade in skins of endangered animals. Though most of this trade is illegal by international treaties, these skins are still occasionally available through evasion of import/export restrictions, on the black market, or by nonparticipating countries. A call to your nearest customs office will usually clarify which hides are considered acceptable. The purchase of such items supports a practice of poaching, and a destructive disregard for and abuse of animal species. The attitude that trade in endangered animal products provides support or economic relief for poorer nations is completely misconceived, as poaching destroys a valuable, living asset.

UPHOLSTERY FILLERS

Fillers for stuffed and padded upholstery are made from a wide range of synthetic and natural fibers. The source of the fiber and its durability are the main environmental concerns. The main health concern is their potential for trapping allergenic dusts.

ANIMAL FIBERS

Animal products, including wool, feathers, down, and hair, are all used as fillings. Except wool, most fillings are the byproducts of the slaughter of animals raised for meat and eggs. True down comes from ducks and geese that are plucked for their feathers up to five times before they are butchered for meat. Of all animal fibers, only wool is likely to be raised in a humane fashion.

About 10%–15% of the population has some degree of allergy to animal-based fibers, particularly feathers and down. In addition, materials may be treated with fire retardants, causing other allergic reactions. Wool is the only exception due to its natural fire retardancy.

PLANT FIBERS

Among plant fibers used for fillings are shredded cotton, kapok, jute, and other tropical fibers. Cotton filling is often low grade cotton of short fiber length which has been formed into "batts." Other plant fibers are used as padding over springs and metal parts.

Plant materials are less allergenic than animal materials, affecting only a small percentage of the population. However, plant fibers are usually treated with fire retardants, which are potential irritants, especially in fillers used as bedding. Recycled materials are generally not allowed in upholstered goods by law in most regions, due to the difficulty in controlling contamination.

SYNTHETIC MATERIALS

The vast majority of upholstered furniture is filled with synthetic foams or spun fibers. Synthetic foams include polyurethane and rubber. (These may also be called polyether, isocyanurate, latex, and synthetic latex foams.) Other plastics, such as polyvinyls, are also used sometimes in foams. All synthetic foams are made from petroleum-based resins which have been expanded with gases to create their cellular structure.

Until recently most high performance (closed-cell type) upholstery foams were made with chlorofluorocarbons (CFCs) which deplete atmospheric ozone when released. As late as 1986, about 36% of the world production of the most ozone depleting CFCs (R11 and R12) were used to make foamed plastics. A great deal of gas is released during foam manufacture, and the remainder leaks slowly from the material over its lifetime (estimated to take 100 years). Toxic methylene chloride was also used in some foaming processes. In most upholstery foams, both these agents have now been replaced with nonozone-depleting alternatives, such as nitrogen and carbon dioxide, which are nontoxic and ozone safe. Some new upholstered furniture bears a seal indicating that the foam is an "ozone safe" type.

The remainder of synthetic fillings are made from the compressed fibers of polyester or acetate. These are similar to the synthetic fibers for textiles discussed earlier.

Virtually all plastic foams are flammable and produce toxic gases during combustion. Urethanes, for example, may produce deadly cyanides when heated. Fire retardants are added to make the foams "self extinguishing" so that they are more difficult to ignite, but toxic gases are still produced when they are heated. The fire retardants are an added problem in foam production because they are also toxic materials. They add environmental risk to the production process, expose industry workers to health hazards, expose upholsterers who cut the material (especially with a "hot wire" process) to health hazards, and add to the likelihood of adverse, allergic reactions by sensitive individuals due to odors or skin contact.

PAINTS AND FINISHES

Most of the major issues about paints are discussed in the paints and coatings section of this book. There are, however, several types which are applied under factory conditions. These provide the most durable and uniform finishes, and are therefore commonly used for furniture.

All factory-applied finishes have the advantage of being handled under conditions that are better controlled than those on a construction site. For example, ventilated spray booths are used to reduce exposure to workers and to control paint loss. Under factory conditions it is possible to control paint storage, mixing, solvent handling, temperature, air flow, air filtration, curing time, and other factors which help to reduce releases of toxic materials into the environment, protect workers from exposure, and produce a better finish.

The best paint shops in furniture plants have recovery systems to minimize the quantity of finish materials lost, very sophisticated ventilation and air filtration systems to protect workers, and stack scrubbers to trap gases and dusts before they are released into the environment. A high quality system can far exceed all occupational health and environmental regulations. The presence of such systems is noted in the Product Reports by the designation of "low emissions plant."

MELAMINES

These are solvent- or water-based acrylic paints modified with butylated melamine and used on metal and wood composition products to produce a tough, washable finish. Prefinished wood shelving is an example of a melamine finish. Dusts from cutting prefinished materials are a health hazard, but the finish itself is chemically stable. (See also PLASTIC LAMINATES, p. 209.)

LACQUERS

Traditional lacquers are solvent-based finishes, usually based on nitrocellulose in a butyl acetate solvent (sometimes containing toluene, xylene and other solvents). They are very hazardous to handle during the period of rapid evaporation of the solvent immediately after application, and can be explosive. There is usually a great deal of hazardous waste produced from lacquer use due to its cleanup requirements. Once cured, however, lacquer-coated products are also chemically stable. Lacquers tend to be brittle and difficult to handle, and are being replaced by safer, water-based finishes.

EPOXIES

These are a large family of resins, usually manufactured as two-part systems with a resin and a catalyst. The production and handling of raw materials for epoxies pose both environmental risks and health hazards to industry workers. The uncured resins and catalysts are highly irritating materials and should not be handled by those without professional experience. The finishes resulting from professionally applied epoxy-based paints are very tough and resist corrosion and stains. They are chemically stable once cured.

SYNTHETIC LACQUERS OR ENAMELS

These are usually acrylics (polymers of acrylic or methacrylic acid esters) or urethanes (isocyanates) which are dispersed in solvents. They are often sold as automotive finishes. A newer formulation called a "compounded prepolymer" is a urethane which is dispersed in water. Somewhat safer to handle than epoxies, these are now popular finishes for high gloss furniture and cabinet surfaces.

BAKED ENAMELS

Many of the above formulations can be heat cured in low temperature ovens to produce a very durable finish. Alkyd resins, epoxies, melamines, and other finishes are cured this way. By definition, any finish which is heat treated above 150°F (66°C) can be classed as a baked finish. The process speeds the release of volatile components of the paint and therefore makes it more chemically stable and safer indoors. The environmental disadvantage of the process is that the gases released by heating are vented to the atmosphere unless sophisticated gas capture and disposal systems are in place.

POWDER COATINGS (ELECTROSTATIC COATINGS)

This is an application method for coating metals which uses no solvents. Metal furniture parts are sprayed with a dry powdered polymer in an electrostatically charged booth. The electrical charge causes the powder to stick to the metal, and it is then heated to fuse it into a hard coating. No solvent is required, and no cleanup is necessary. Less than 2% of the finish material is wasted. Conventional spray processes may waste over 40%. The environmental and health benefits of this system are significant because both toxic waste and worker exposure to solvent is eliminated. The process can even be used to replace plated, protective coatings like cadmium, which have serious environmental and health consequences.

DURABILITY AND RECYCLING

DURABILITY

Good quality furniture will last several generations and increase in value with time. While carpets, draperies, wall coverings and other short-lived interior materials may last only a few years and then are discarded in landfills, furniture should be selected to endure. Up to 28% of the volume of urban landfills is waste demolition and construction materials. If the correct choices are made, furniture need never end up there.

REUSE

Many businesses find good value in used office furniture. With current high office turnover rates, there is a large inventory available either from brokers, or from the original manufacturers who buy back their own products for reconditioning and resale. Most high quality office furniture is durable and can be reconditioned for a modest cost, particularly classic, functional designs, a sometimes unrecognized element of durability.

RECYCLING

When furniture or furniture parts are no longer usable, some of the materials are recyclable. Nearly all metal parts can be salvaged for resmelting, padding foams and fillings can be shredded for packaging material, and glass parts can be recycled. The parts that currently have no recycling capability are most plastic, leather, fabric and wood parts. For all composite materials or products, easy disassembly increases the potential for recycling, and some manufacturers are incorporating this philosophy in the design of the product. Furniture made from recycled plastic is now available, though it is generally limited to garden furniture. The mixed plastics used in its manufacture make a low grade but durable product, which helps to reduce landfill waste.

MANUFACTURER

Gesika (Germany)
Imported by Gesika Furniture Inc.
155A Matheson Blvd. West
Mississauga, ON L5R 3L5
Tel (416) 507-0184   Fax (416) 507-4529

RELATED PRODUCTS

GESIKA SYSTEMS FURNITURE
GESIKA PANEL SYSTEM

DESCRIPTION

GESIKA STORAGE AND STORAGE WALL RANGE: Clean, contemporary, and completely modular styling for any office decor. The system is comprised of credenzas, cupboards with sliding doors, card index and filing cabinets, wardrobes, add-on cupboards, roller blind cupboards, add-on cabinets, kitchen cupboards, and doors in a wide variety of widths and heights.

There is a wide variety of media storage accessories (brackets, pullout racks, shelves, etc.).

All cupboards have rear panel holders as a standard. Decorative panels can be fitted quickly and easily at any time, allowing Storage Wall to act as a furniture/panel system.

MANUFACTURER'S STATEMENT

Gesika's manufacturing plant collects wood dust and waste chips for a waste-to-energy scheme that heats the plant in winter months. The boiler system has a gas purification system which ensures that no toxic fumes are released into the air. The particle board used in the production of all furniture items is a low formaldehyde emitting type. Acrylonitrile butadiene styrene coatings are used instead of polyvinyl chloride, releasing less toxic fumes on incineration. Only water-based polyvinyl acetate glues are used.

Storage wall system features include its verstile configuration, ease of mounting, organizational versatility, high stability, and the reusability of individual cupboards. The room dividing capacity (floor to ceiling, or freestanding) eliminates the necessity to construct permanent walls and built-in cabinetry, and thus the resulting construction waste when the time comes for office expansion or relocation.

Gesika has addressed some of the environmental ramifications of office relocation and expansion, given that the statistics show that companies relocate every 3 years. The versatility benefits of Gesika made the product comparable in price (considering materials, labor, demolition and waste) to conventional office construction, but without the demolition waste materials to be disposed of, or the depletion of new materials for reconstruction.

PRODUCTION

IN-PLANT
ENERGY
EFFICIENCY
& RECYCLING

LOW
EMISSIONS
PLANT

PACKING / SHIPPING

INSTALLATION / USE

LOW TOXIC
EMISSIONS
IN USE

DURABLE

SIMPLE,
NONTOXIC
MAINTENANCE

RESOURCE RECOVERY

REUSABLE,
SALVAGEABLE

OTHER

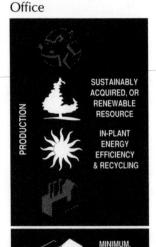

PRODUCT
Enclosed and open office furniture

MANUFACTURER

Herman Miller, Inc.
8500 Byron Rd.
Zeeland, MI 49464
Tel (616) 772-5323, (800) 851-1196
Fax (616) 772-4129

Herman Miller Canada, Inc.
11 Adelaide St. W., Suite 400
Toronto, ON M5H 3X9
Tel (416) 366-3300
Fax (416) 366-2100

DESCRIPTION

THE SANFORD™ COLLECTION: Provides a range of looks from traditional to contemporary. Unique wire management system within desks, credenzas, return and bridges (optional). Available with options of finishes, edge details, and pull finishes.

BURDICK GROUP™ FURNITURE: Designed by Bruce Burdick. Components include polished aluminum beams to support surfaces. System includes trays, dividers, drawers, machine tables, project shelves, and top surfaces. Options for top surfaces are wood, marble, glass, laminate.

NEWHOUSE GROUP® FURNITURE: Table desks, pedestal desks, credenzas, VDT tables, cabinets, and lateral files combine contemporary lines and modern materials. Won the Institute of Business Designers Gold Award in 1987.

RELAY® FURNITURE: Designed by Geoff Hollington, line includes tables, desks, credenzas, bookcases, mobile storage carts, work organizers, and screens.

ACTION OFFICE FURNITURE SYSTEM: Full office furniture system, including panels, work surfaces, storage units.

ETHOSPACE OFFICE FURNITURE SYSTEM: Full office furniture system, including panels, work surfaces, storage units.

MANUFACTURER'S STATEMENT

Herman Miller has developed a comprehensive design strategy to monitor and minimize its impact on the environment, and factoring the environmental and ethical costs into the corporate bottom line. Herman Miller has formulated a company-wide environmental policy and acknowledged that some classic pieces required modification to conform to that policy. The company is committed to using only woods from sources that implement sustainable forest management, and has engaged Rainforest Action Network as a consultant to determine policy parameters.

Other components of the policy include recycling initiatives, reduction of packaging and shipping materials, distribution of company mugs to employees, review of the impact of acquisition of source materials, activities of the company's submanufacturers, reduction of waste, and design of products for disassembly and remanufacture.

A waste-to-energy plant, for the incineration of benign waste materials, was constructed, and since 1982 the cogenerator supplies all heating and air conditioning needs at the central plant, as well as producing incidental amounts of electricity. The byproduct wood ash is sold to a local cinder block manufacturer, eliminating high landfill and energy costs.

PRODUCT
Open office furniture systems

MANUFACTURER

Knoll North America Corp.
655 Madison Ave.
New York, NY 10021
Tel (800) 445-5045

The Knoll Group
100 Arrow Rd.
Toronto, ON M9M 2Y7
Tel (416) 741-5453
Fax (416) 741-1494

DESCRIPTION

REFF SYSTEM 6: Executive free-standing system, with wood or laminate work surfaces and wood interior. The Reff System panel has replaceable panel skins, allowing resurfacing, without the need to purchase an entire new panel.

REFF SYSTEM Z, MORRISON NETWORK, EQUITY: Open office system, with upholstered panels, laminate work surfaces, and metal interior.

KNOLLEXTRA: Extensive line of paper management and desk accessories.

KNOLLSTUDIO: Classic furniture, lounge seating renowned for designs by illustrious designers and architects.

BULLDOG CHAIR: Conceived at all stages for clean technology. Hot-melt, VOC-free glues are used to fasten freon-free foam to the chair's shell. Chair's components are recyclable. Ergonomically designed.

MANUFACTURER'S STATEMENT

Product life is extended, eliminating wasteful frequent replacement, by making available maintenance information and replaceable parts to customers. 15-year warranty is provided to the initial user.

A "cradle-to-grave" approach includes utilizing recycled materials in products that can easily be disassembled for further recycling. The imprinting of all plastic parts with the appropriate recycling symbols is in the planning.

In the finishing plant, by adopting new processes and state-of-the-art technology, Knoll has achieved a 6-fold reduction in VOCs, based on product volume, since 1983. It has surpassed U.S. Federal guidelines for reducing VOCs by 10%, and improvements are ongoing. Waste office paper and plant scraps recycling programs are in effect. The plant is self-sufficient in heat requirements. Minimal packaging waste is generated, with returnable, reusable poly bags or blanket wrapping.

The Knoll Group has joined forces with Green Cross to establish a certification process for producers and manufacturers of wood. They are currently investigating sustainable sources for obeche, mahogany, maple, oak, and cherry.

PRODUCTION

SUSTAINABLY ACQUIRED, OR RENEWABLE RESOURCE

IN-PLANT ENERGY EFFICIENCY & RECYCLING

PACKING / SHIPPING

MINIMUM, RECYCLED, RECYCLABLE PACKAGING

INSTALLATION USE

DURABLE

SIMPLE, NONTOXIC MAINTENANCE

RESOURCE RECOVERY

REUSABLE, SALVAGEABLE

OTHER

FAIR BUSINESS PRACTICES

RESEARCH & EDUCATION PROGRAMS

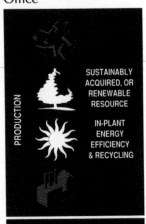

PRODUCTION

SUSTAINABLY
ACQUIRED, OR
RENEWABLE
RESOURCE

IN-PLANT
ENERGY
EFFICIENCY
& RECYCLING

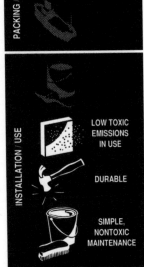

PACKING / SHIPPING

MINIMUM,
RECYCLED,
RECYCLABLE
PACKAGING

INSTALLATION / USE

LOW TOXIC
EMISSIONS
IN USE

DURABLE

SIMPLE,
NONTOXIC
MAINTENANCE

RESOURCE RECOVERY

OTHER

PRODUCT
Office and residential furniture

MANUFACTURER

Nienkamper Furniture
    & Accessories
415 Finchdene Square
Scarborough, ON M1X 1B7
Tel (416) 298-5700
Fax (416) 298-9535

L.A.: Tel (310) 659-0733
    Fax (310) 652-7068
Dallas: Tel (214) 352-8262
    Fax (214) 352-8040
Chicago: Tel (312) 222-1336
    Fax (312) 664-9844
New York: Tel (212) 475-8575
    Fax (212) 489-0073

DESCRIPTION

LOUNGE SEATING COLLECTIONS: All frames are made from soft domestic maple, and CFC-free upholstery foams (by Carpenter).

CASE GOODS AND FREE-STANDING OFFICE FURNISHINGS: "Fibrewood" MDF, a low formaldehyde emitting product is used, and fully sealed with finish, thereby further reducing possible emissions. It is used as a decorative material, replacing the need for exotic veneers, with a wide range of semi-transparent finishes.

All lacquers used are catalytically cured, and once cured, are inert. Curing takes place in the manufacturing facility.

DESK ACCESSORIES: Leather cutoffs are salvaged from the sofa manufacturing process.

MANUFACTURER'S STATEMENT

Solid wood cutoffs are used as firewood or recycled, upholstery foam is collected and sold for conversion into foam undercushion.

80% of product is blanket wrapped for shipping, and the rest is packaged with recyclable cardboard, wood, and reused foam.

Nienkamper is committed to recycling and reuse of materials, and the replacement of materials and processes as technological and materials advances are made, and the promotion of these materials to the industry and design community.

PRODUCT
Remanufactured office furniture

MANUFACTURER

Phoenix Designs (subsidiary of Herman Miller)
10875 Chicago Dr.
Zeeland, MI 49464
Tel 1 (800) 253-2733, (616) 772-5323   Fax (616) 772-4129

DESCRIPTION

ASNEW™ FURNITURE SYSTEM: Remanufactured version of Action Office, Herman Miller's best-selling systems furniture.

Available in all Action Office finishes and the most popular fabrics. As-New™ components work with Action Office, and fabrics and trims match. AsNew™ has the Underwriters Laboratory label, and has Class A fire rating. Furniture can be rented, leased, or bought. Trade-ins on old Action Office components go toward the purchase of an AsNew™ system.

Available in two weeks.

MANUFACTURER'S STATEMENT

Action Office components are taken apart. The hardware and metal parts are sanded and repainted. New plastic laminate and T-moulding are installed. Old panel fabric is removed and replaced.

See also p. 220.

PRODUCTION

RECYCLED CONTENT

IN-PLANT ENERGY EFFICIENCY & RECYCLING

LOW EMISSIONS PLANT

PACKING / SHIPPING

INSTALLATION / USE

DURABLE

SIMPLE, NONTOXIC MAINTENANCE

RESOURCE RECOVERY

REUSABLE, SALVAGEABLE

OTHER

FAIR BUSINESS PRACTICES

RESEARCH & EDUCATION PROGRAMS

PRODUCTION

SUSTAINABLY
ACQUIRED, OR
RENEWABLE
RESOURCE

IN-PLANT
ENERGY
EFFICIENCY
& RECYCLING

LOW
EMISSIONS
PLANT

PACKING / SHIPPING

MINIMUM,
RECYCLED,
RECYCLABLE
PACKAGING

INSTALLATION / USE

DURABLE

RESOURCE RECOVERY

OTHER

FAIR
BUSINESS
PRACTICES

RESEARCH &
EDUCATION
PROGRAMS

PRODUCT
Adjustable office task seating and occasional seating

MANUFACTURER

Steelcase Inc.
P.O. Box 1967
Grand Rapids, MI 49501
Tel (616) 247-2710

Steelcase Canada Ltd.
7200 Woodbine Ave.
Markham, ON L3R 1A2
Tel (416) 475-6333
Fax (416) 475-3653

DESCRIPTION

RALLY, CRITERION SERIES, SENSOR CHAIR LINE: Fully adjustable, ergonomically designed.

TRILOGY™ SERIES, 430 AND 421 SERIES: Functional, comfortable, versatile, fully adjustable arm and task chairs.

CONCENTRX AND 454 SERIES: Fully adjustable task chairs dissipate static electricity charges and prevent further buildup.

PARADE SERIES, AND 472 SERIES: Stacking, ganging, plastic chairs.

CORVO™, SNODGRASS COLLECTION: Solid, classic style side/guest chair, designed by Orlando Diaz-Azcuy. Wood frame, upholstered back and seat. Classic, versatile side/guest chair designed by Warren Snodgrass, available with wood or metal frames.

MANUFACTURER'S STATEMENT

Steelcase employs 3 environmental engineers, and 3 industrial hygienists. The office has voluntarily conducted audits at all manufacturing facilities and subsidiaries, since 1987. Steelcase has received awards from governmental agencies for a habitat restoration project, and advanced air pollution control.

Since 1978, a VOC emission reduction program has been in effect, and a new state-of-the-art emission control system will reduce solvent emissions to more than 60% below current legislated levels. New equipment and practices continually reduce solid and gas emissions, through replacement of toxic materials and recycling, and waste-to-energy schemes.

Packaging and shipping materials have been reduced by as much as 60% by substituting biodegradable stretchwrap for traditional corrugated cartoning. 50% of Sensor chairs are shipped with reusable polyurethane foam buns. 35% of other products are shipped blanket-wrapped.

Steelcase recycles solvents, plating chemicals, and plastics, as well as the 5 million lbs. of scrap steel produced per month. Office paper recycling programs are in effect.

Employees actively support such company initiatives as car pooling, recycling efforts, and participating in community recycling events.

Since 1983, the company, with *Interiors* magazine, has sponsored annual environmental symposia, dealing with a wide array of topics. Steelcase environmental engineers serve on community task forces.

PRODUCT
Office furniture systems, and filing and storage cabinets

MANUFACTURER

Steelcase Inc.
P.O. Box 1967
Grand Rapids, MI 49501
Tel (616) 247-2710

Steelcase Canada Ltd.
7200 Woodbine Ave.
Markham, ON L3R 1A2
Tel (416) 475-6333
Fax (416) 475-3653

RELATED PRODUCTS

WORKFLO: Rail-based system of work tools and surfaces, and accessory objects for systems furniture. Designed by the Richard Penney Group.

DESCRIPTION

AVENIR™ SYSTEMS FURNITURE: A system that has grown from Steelcase's Movable Walls system, allowing components purchased in the past to be incorporated into the overall group of furniture. Evolved over 20 years, includes panels, work surfaces, storage units, storage accessories, and a lighting system.

9000 SERIES: Freestanding systems furniture, not requiring to be fixed to panels. Unusual components for a nontraditional systems look. Includes panel, desk, work surfaces, chairs, and file cabinets. Wide choice of fabrics, laminates, and paint colors.

BASIX (SERIES 9000): Modular, functional work stations, including panels, work surfaces, and filing/storage units.

800 AND 900 SERIES, VERTICAL FILES: Lateral files and storage cabinets, available in custom color paint finishes, with options of laminate or wood tops, side panels. Fast shipping and wide choice of sizes.

ULTRONIC 9000, ELECTRONIC SUPPORT FURNITURE: Systems furniture for the electronic office. Ergonomic, flexible, adjustable, includes panels, work surfaces, storage, VDT and computer stands, mobile stands, media storage, ergonomic devices, storage pedestals, computer accessories.

MANUFACTURER'S STATEMENT
See p. 224.

PRODUCTION — SUSTAINABLY ACQUIRED, OR RENEWABLE RESOURCE · IN-PLANT ENERGY EFFICIENCY & RECYCLING · LOW EMISSIONS PLANT

PACKING / SHIPPING — MINIMUM, RECYCLED, RECYCLABLE PACKAGING

INSTALLATION / USE — DURABLE · SIMPLE, NONTOXIC MAINTENANCE

RESOURCE RECOVERY — REUSABLE, SALVAGEABLE

OTHER — FAIR BUSINESS PRACTICES · RESEARCH & EDUCATION PROGRAMS

PRODUCTION

SUSTAINABLY ACQUIRED, OR RENEWABLE RESOURCE

IN-PLANT ENERGY EFFICIENCY & RECYCLING

LOW EMISSIONS PLANT

PACKING / SHIPPING

MINIMUM, RECYCLED, RECYCLABLE PACKAGING

INSTALLATION / USE

DURABLE

SIMPLE, NONTOXIC MAINTENANCE

RESOURCE RECOVERY

REUSABLE, SALVAGEABLE

OTHER

FAIR BUSINESS PRACTICES

RESEARCH & EDUCATION PROGRAMS

PRODUCT
Wood office furniture

MANUFACTURER

Steelcase Wood Furniture
(formerly Stowe & Davis)
4300 44th St.
Kentwood, MI 49508
Tel (616) 698-5735

Steelcase Canada Ltd.
37 Front St.
Toronto, ON M5E 1B3
Tel (416) 366-3300
Fax (416) 366-2100

DESCRIPTION

NEO® SYSTEM: Corporate, executive systems furniture. Wood body and wood or laminate work surfaces, fabric panels with wood trim, storage and cable/wire handling feature.

EDGEWOOD™: Contemporary line of fine wood furniture designed by Robert Taylor Whalen. Wide variety of veneer and accents (leather, chrome, bronze). Line includes storage walls, desks, table, and work stations.

MANUFACTURER'S STATEMENT

Stowe & Davis, Steelcase's wood division, has a Forest Preservation Policy, stating that it will obtain wood only from sustained yield programs. It is committed to identifying, developing, and offering creative finishes, wood alternatives, and nonendangered woods from domestic, sustained yield sources.

In regard to other environmental responsibility and action, Stowe & Davis follows Steelcase's corporate lead.

PRODUCT
Complete office furniture system

MANUFACTURER

Teknion Furniture Systems Inc.
1150 Flint Rd.
Downsview, ON M3J 2J5
Tel (416) 661-3370
Fax (416) 661-5137

Teknion Inc.
17 W. Stow Rd., P.O. Box 562
Marlton, NJ 08053
Tel (609) 596-7608
Fax (609) 596-8088

DESCRIPTION

Complete office furniture system that incorporates stacking panel, and power and telecommunications access at work surface height with design flexibility. Steel panel frames support fabric acoustic facings, with laminate and wood work surfaces. Case goods in laminate and wood, and steel storage units are available.

Teknion's recut wood veneer, called *Flintwood,* is produced from a fast growing, cultivated and managed African tree. Obeche matures in 30–40 years. One tree is cut for every 2.2 acres of forest. Peeled veneer is pressure aniline dyed, for consistent color, and then laminated together and recut into veneer or solid members producing a harder veneer, with the possibility of varying types of wood grain patterns: straight grain, cathedrals, burl, and bird's eye.

All metal surfaces are coated with epoxy polyester powder, a solvent-free process which utilizes 98% of the material. The finish is durable, more abrasion resistant than conventional coatings, with less danger of corrosion of the substrate.

The system's TU100 lighting is an energy-efficient 13-watt tube, used with a high-quality reflector. Less heat is generated, which reduces air conditioning requirements.

MANUFACTURER'S STATEMENT

Flintwood veneers allows 100% use of the raw material, while reconstituting the wood to exacting specifications maintains consistency of color and grain patterning.

Cardboard packaging has been minimized, and where possible, replaced by blanket or scrap fabric wrapping. Less packaging with the use of stretch-wrap film reduces damage claims, as shippers are more careful handling a product they can see. Teknion has reduced packaging materials by 500,000 lbs. (cardboard) and 65,000 lbs. (plastic) per year. All scrap steel, aluminum, and cardboard are recycled. Fabric remnants and scraps are used as packaging or sent to schools for use in art programs.

State-of-the-art manufacturing plant is energy efficient, and recycles water continually. The plant's heating and air conditioning systems are state of the art. A comprehensive recycling program is in effect, which includes office paper, cans, bottles, printer cartridges, etc. Company mugs are distributed to all staff.

Environmental audits and assessments are conducted periodically within the plant and office, by both specialized auditors and members of the Teknion Health and Safety Committee.

PRODUCTION
- SUSTAINABLY ACQUIRED, OR RENEWABLE RESOURCE
- IN-PLANT ENERGY EFFICIENCY & RECYCLING
- LOW EMISSIONS PLANT

PACKING / SHIPPING
- MINIMUM, RECYCLED, RECYCLABLE PACKAGING

INSTALLATION / USE
- LOW TOXIC EMISSIONS IN USE
- DURABLE
- SIMPLE, NONTOXIC MAINTENANCE

RESOURCE RECOVERY
- REUSABLE, SALVAGEABLE

OTHER
- FAIR BUSINESS PRACTICES

PRODUCTION

SUSTAINABLY
ACQUIRED, OR
RENEWABLE
RESOURCE

PACKING SHIPPING

MINIMUM,
RECYCLED,
RECYCLABLE
PACKAGING

INSTALLATION USE

LOW TOXIC
EMISSIONS
IN USE

DURABLE

SIMPLE,
NONTOXIC
MAINTENANCE

RESOURCE RECOVERY

REUSABLE,
SALVAGEABLE

OTHER

FAIR
BUSINESS
PRACTICES

PRODUCT
Commercial and residential quality teakwood outdoor furniture and site furnishings

MANUFACTURER
Barlow Tyrie Inc. (England)
1263-230 Glen Ave.
Moorestown, NJ 08057
Tel (609) 273-7878, (800) 451-7467   Fax (609) 273-9199

RELATED PRODUCTS
Stains, pp. 161–65

DESCRIPTION
Full range of outdoor furniture including: benches (with and without cotton cushions), chairs and bar stools, tree seats, loungers, round and square tables, trolleys, folding chairs and tables, planters, and parasols.

The understated elegance of traditional Edwardian design detail, quality craftsmanship and construction methods are the main features of the furniture line. Pieces are natural and unfinished and require no finish for outdoor use.

MANUFACTURER'S STATEMENT
Barlow Tyrie has been awarded the Good Wood Seal of Approval by Friends of the Earth, for buying teak exclusively from strictly controlled plantations in Java (Indonesia). Once harvested, the land is replanted by local farmers who tend the seedlings and are permitted to underplant with their own food crops.

Barlow Tyrie has entered into a joint venture with the government of Indonesia which will create an estimated 30–60 new jobs, where one meal per day will be provided by the company.

PRODUCT
Residential teak furniture, for outdoor use

MANUFACTURER
Lister by Geebro (England)
1900 The Exchange, Suite 655
Atlanta, GA 30339
Tel (404) 952-4272   Fax (404) 952-4489

DESCRIPTION
A full range of traditional and contemporary handmade outdoor teak garden furniture: seats, table, benches, planters, and loungers in 9 styles.

MANUFACTURER'S STATEMENT
Geebro has a history of craftsmanship dating back more than 150 years. After its acquisition in 1978, the Lister range of garden and leisure furniture has been manufactured in the market town of Hailsham.

Lister outdoor furniture is built from solid lengths of teak. Alert to its responsibilities, the company uses only farmed teak, taken from the state controlled plantations in Java, Indonesia. For every tree taken, another is planted. The continuation of the forest and income of that country is assured. The timber used is of the highest quality, and boasts a distinctive, warm color and grain. Only mature trees are selected and felling doesn't occur until the teak is 80 years old. Once felled, the wood is taken by ship to England and crafted by traditional methods: mortice and tenon joinery, and traditional and contemporary designs.

In 1989, the Lister line was awarded the Friends of the Earth Seal of Approval, and listing in the Good Wood Guide.

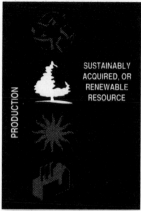

PRODUCTION

SUSTAINABLY ACQUIRED, OR RENEWABLE RESOURCE

PACKING / SHIPPING

INSTALLATION / USE

LOW TOXIC EMISSIONS IN USE

DURABLE

SIMPLE, NONTOXIC MAINTENANCE

RESOURCE RECOVERY

REUSABLE, SALVAGEABLE

OTHER

PRODUCTION

SUSTAINABLY
ACQUIRED, OR
RENEWABLE
RESOURCE

IN-PLANT
ENERGY
EFFICIENCY
& RECYCLING

PACKING / SHIPPING

MINIMUM,
RECYCLED,
RECYCLABLE
PACKAGING

INSTALLATION / USE

LOW TOXIC
EMISSIONS
IN USE

DURABLE

SIMPLE,
NONTOXIC
MAINTENANCE

RESOURCE RECOVERY

OTHER

FAIR
BUSINESS
PRACTICES

PRODUCT
Nontoxic residential furniture (mail-order)

MANUFACTURER
Pure Podunk, Inc.
RR#1, Box 69,
Thetford Center, VT 05075
Tel (802) 333-4505

RELATED PRODUCTS
Auro finishing products, pp. 145, 166
Livos finishing products, pp. 147, 150, 161, 168, 185

DESCRIPTION
"BEST OF VERMONT" COLLECTION: Selection of pieces from Vermont wood shops manufacturing furniture from cherry, maple, oak, and ash from sustainable sources. All items are unfinished. If they contain glues, they are formaldehyde-free, no-odor white or yellow glues. Some of the beds come with cloth straps to hold the slats in place; some of the beds use softwood slats to support the mattress.

Furniture items include night tables/bedside stands, available in different heights. Bedframes come in an extensive selection of styles, some with headboard and footboard, headboard only, or four-poster canopy. Most are available in twin, double, queen, and king sizes. Call for separate bedframe catalog/flyer.

Several styles of convertible couch frames and daybeds are available in "L" or bifold styles, with or without arms, for use with futons or cushions. Range of sizes from twin, double, and queen. Some have matching armchairs. Cushions are stuffed with organic wool and cotton, and covered with 100% untreated, prewashed cotton fabric. Woods include cherry, white ash, or maple. Available with an optional clear *Livos* finish.

Ottomans, coffee tables, end tables, behind-the-couch tables, and parson's tables are available in cherry or maple, in different sizes, some rectangular or oval shapes.

MANUFACTURER'S STATEMENT
Pure Podunk has a passionate concern for personal and environmental health, and a desire to see vital, sustainable rural communities. Pure Podunk is committed to the revival of agragarian economy and culture, by supporting local nonchemical farms, sustainable yield logging operations, and operating as a rural cottage industry.

Pure Podunk donates 7% of its profits to charitable organizations concerned with these issues.

See also p. 241.

PRODUCT
Commercial and residential teak and mahogany furniture, for indoor or outdoor use

MANUFACTURER
Summit Furniture, Inc.
P.O. Box S
Carmel, CA 93921
Tel (408) 394-4401
Fax (408) 394-5242

RELATED PRODUCTS
Stains, pp. 161–65

DESCRIPTION
A full range of bold, contemporary handmade indoor and outdoor teak and mahogany furniture designs: sofas, lounge chairs and adjustable chaises, ottomans, benches, dining chairs, bar stools, tables, servers, bar carts, planters, and garden lights.

Hand-rubbed light oil and pickled white finishes are available for outdoor use. Other hand-rubbed finishes available for indoor use. Parts are joined, glued, and secured with brass screws and hardware.

MANUFACTURER'S STATEMENT
Summit's sole supply of teak and mahogany for the past 10 years is from Indonesian plantations, certified as Smart Wood by the Rainforest Alliance.

Offcuts and scraps of wood are utilized in other projects. Warehouse and offices are new, energy efficient construction.

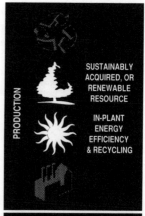

PRODUCTION

SUSTAINABLY ACQUIRED, OR RENEWABLE RESOURCE

IN-PLANT ENERGY EFFICIENCY & RECYCLING

PACKING / SHIPPING

MINIMUM, RECYCLED, RECYCLABLE PACKAGING

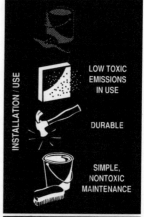

INSTALLATION / USE

LOW TOXIC EMISSIONS IN USE

DURABLE

SIMPLE, NONTOXIC MAINTENANCE

RESOURCE RECOVERY

REUSABLE, SALVAGEABLE

OTHER

PRODUCTION

SUSTAINABLY
ACQUIRED, OR
RENEWABLE
RESOURCE

PACKING / SHIPPING

INSTALLATION / USE

LOW TOXIC
EMISSIONS
IN USE

DURABLE

SIMPLE,
NONTOXIC
MAINTENANCE

RESOURCE RECOVERY

OTHER

PRODUCT
Pure fiber residential fabrics (mail-order)

MANUFACTURER
The Cotton Place
P. O. Box 59721
Dallas, TX 75229
Tel (214) 243-4149, (800) 451-8866

DESCRIPTION
NATURALGUARD BARRIER CLOTH™: 100% cotton, custom manufactured and contains no permanent chemical finish. Made from woven pima cotton, with 280 thread count. The fine weave allows it to act as a protective barrier against dust, pollen, and other foreign particles found in pillows, mattresses, upholstery, and auto interiors. No unnecessary chemicals are used in the manufacture: it is designed to provide protection from chemical and environmental irritants. Ordinary barrier cloth has a chemical finish that will withstand only 100 washings. 45" wide, available in white only.

PIMA BROADCLOTH: 100% cotton, 45" wide, extra-fine, long staple cotton from England. Available in white and 12 colors.

CHEESECLOTH: 100% cotton, 36" wide, high quality, bleached.

CORDUROY: 100% cotton, 7 and 11 wale, 45" wide. Available in 5 colors.

DUCK: 100% cotton, 36" wide, 8 oz., bleached.

FELT PADDING: 100% cotton, 72" wide, unbleached, woven, double-faced, 13 oz. (similar to heavy flannel).

MUSLIN: 100% cotton, 140 thread count per inch, 72" wide. Available in bleached, unbleached, and medium blue.

LOUVAIN LINEN: 100% linen from Belgium. Long wearing, machine wash and dry. 72" wide, available in white only. 42% linen/58% cotton, 60" wide. Available in white only.

SILK CHARMEUSE: Crepe-backed silk satin, 45" wide, available in ivory and white only.

SILK CREPE DE CHINE: 100% silk, 45" wide. Available in white and 2 colors.

CHINA SILK: 100% silk, 45" wide. Available in white and 8 colors.

NOIL: 100% silk, 36" wide. Available in natural and 7 colors.

MANUFACTURER'S STATEMENT
The Cotton Place specializes in products suitable for persons with chemical sensitivities.

MANUFACTURER

Guilford of Maine, Inc.
(an Interface Company)
Oak St.
Guilford, ME 04443
Tel (207) 876-3331
Fax (207) 876-3538

Guilford of Maine
(Canada) Inc.
254 St-Urbain
Granby, PQ J2G 8M8
Tel (514) 777-3411

DESCRIPTION

PANEL FABRICS: 100% polyester fabrics in a wide range of colors and weaves, 66" wide. Patterns include: *FR 701®*, *FR 701® Hobnail*, *Carina*, *Frise*, *Belgrade*, and *Crosscurrents*.

Blended fiber fabrics in a wide range of colors and weaves, 66" wide. Patterns include: *Donegal Tweed*, *Saxony Flannel*, and *Saxony Pindot* (polyester/wool), and *Tussah Silk* (polyester/silk/flax).

UPHOLSTERY FABRICS: Synthetic fabrics in a wide range of colors and weaves, 54" wide, with latex backing. Patterns include: *Paradiso* and *Biella* (polyester), *Willington*, *Byram Riverside*, *Audubon*, *Belle Haven* (polypropylene—contains Intersept®), *Spoletto*, *Sumi* (wool/nylon), and *Phoenix* and *Pavo* (nylon/modacrylic).

MANUFACTURER'S STATEMENT

Guilford of Maine's corporate goal is to review areas where waste material or byproducts can be reduced, eliminated, recycled, or reused; to review and adjust as required their environmental programs; maintain a staff and management of environmentally conscientious professionals to direct the Interactive Partnership program.

Upon approval, Guilford of Maine intends to compost and landspread of their treatment plant residues and wood ash from boilers. Their current Environmental Preservation programs include in-office and in-plant recycling of paper, cardboard, scrap metal, plastic bags for fabric, waste hydraulic oils, and solvents; reuse of fabric wall covering from renovation projects, fabric remnants as retail goods, and scrap yarn on secondary markets. Waste wood from a local mill now supplies boilers, providing 99% of plant energy.

Guilford of Maine has introduced contract fabrics treated with Intersept®, a biocide which is licensed by the Environmental Protection Agency. Guilford of Maine is a partner of the Envirosense™ consortium.

NOTE

The addition of biocides to materials may not be an appropriate practice for many applications. Biocides are not a substitute for adequate cleaning and maintenance.

PRODUCTION — IN-PLANT ENERGY EFFICIENCY & RECYCLING / LOW EMISSIONS PLANT

PACKING / SHIPPING — MINIMUM, RECYCLED, RECYCLABLE PACKAGING

INSTALLATION / USE

RESOURCE RECOVERY

OTHER — RESEARCH & EDUCATION PROGRAMS

PRODUCTION
SUSTAINABLY
ACQUIRED, OR
RENEWABLE
RESOURCE

LOW
EMISSIONS
PLANT

PACKING / SHIPPING

INSTALLATION / USE
LOW TOXIC
EMISSIONS
IN USE

DURABLE

SIMPLE,
NONTOXIC
MAINTENANCE

RESOURCE RECOVERY

OTHER

RESEARCH &
EDUCATION
PROGRAMS

PRODUCT
Naturally colored cotton (without dyes)

MANUFACTURER
Natural Cotton Colours, Inc.
P. O. Box 791,
Wasco, CA 93280

DESCRIPTION
FOXFIBRE YARN: Organic, naturally colored cotton yarn available in 3 weights:

4,200 yds./lb., in solid colors: coyote (brown), green, camel (tan); and mixed colors: coyote/white, green/white.

1,680 yds./lb., in solid colors: coyote (brown), green.

840 yds./lb., in solid colors: coyote (brown), green, camel (tan); and mixed colors: coyote/camel, camel/coyote, coyote/camel/white, white/camel/coyote, green/white, white/green/coyote.

FOXFIBRE SLIVER: Coyote/Camel Breaker Sliver has been carded twice. When spun and boiled, coyote darkens, but camel remains unchanged, creating a heathered effect.

Green Card Sliver has been carded once and is similar to handcarded cotton in its preparation. Green is susceptible to fading unless it is spun and boiled. Boiling darkens the color, depending on the water's mineral content and pH level.

Breeder's Mix Card Sliver is a combination of all the remaining cotton plants after all selections are made from the breeding nursery. The final slivers are a khaki color.

Coyote Card Sliver is the closest to handcarded of all the sliver types. The longer the sliver boils, the darker it will become.

Acala Card Sliver is grown in San Joaquin Valley and is a bright white.

Organically Grown Lint coyote and breeder's mix.

MANUFACTURER'S STATEMENT
While attempting to breed higher quality into wild brown cottons, so that they could be spun with modern equipment, fibers which grow in several colors (green, red, and pink) were discovered . The colors take on peculiarities of tone and shade unique to season, climate, and soil, and they can intensify with washing. It is believed that ancient domesticators of cotton in the New World developed a wide variety of naturally colored cotton.

All the cotton from the breeding nursery is pesticide free, and organic methods suitable for large-scale production systems are in development.

PRODUCT
Polyurethane foam cushioning

MANUFACTURER

E. R. Carpenter Company, Inc.
P.O. Box 27205
5016 Monument Ave.
Richmond, VA 23230
Tel (804) 359-0800
Fax (804) 354-0858

DESCRIPTION

QUALUX®: Polyurethane foam cushioning for furniture seats and backs, foam mattress units, and mattress toppers. Available in 8 grades.

HR LUX™: Polyurethane foam cushioning where high resiliency performance is required. Available in 4 grades.

OMALUX® DENSIFIED FOAM: Polyurethane foam cushioning for furniture seats and backs, all foam mattress units and topper, pads and automotive seating. Available in 8 grades.

MANUFACTURER'S STATEMENT

E. R. Carpenter Company has been producing high quality, high density, CFC-free foams since 1971. No auxiliary blowing agents of any kind are used.

FURNITURE
Fabrics

PRODUCTION

LOW EMISSIONS PLANT

PACKING / SHIPPING

INSTALLATION / USE

LOW TOXIC EMISSIONS IN USE

RESOURCE RECOVERY

OTHER

235

PRODUCTION

LOW
EMISSIONS
PLANT

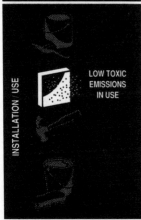

PACKING / SHIPPING

INSTALLATION / USE

LOW TOXIC
EMISSIONS
IN USE

RESOURCE RECOVERY

OTHER

PRODUCT
Polyurethane foam cushioning

MANUFACTURER

Union Carbide
Specialty Chemicals Division
39 Old Ridgebury Rd.
Danbury, CT 06817-0001

Union Carbide
ACWV Polyol Company
Ste. 210, 1210 Sheppard Ave. E.
Willowdale, ON M2K 1E3
Tel (416) 488-4189

DESCRIPTION

ULTRACEL FOAM: Urethane foam cushioning for home furnishing applications.

MANUFACTURER'S STATEMENT

Union Carbide has developed the first process to use no auxiliary blowing agent, such as CFC or methylene chloride.

PRODUCT
Foreign, handcrafted accessories

MANUFACTURER
Various artisans and cooperative groups worldwide
Imported by Bridgehead (Oxfam Canada)
424 Parkdale Ave.
Ottawa, ON K1Y 9Z9
Tel (613) 567-1455   Fax (613) 567-1468

DESCRIPTION
Wide variety of handcrafted personal and household accessories, including: cushion and duvet covers, woven wall hangings, desk accessories, candlesticks, shelving units, tribal artifacts, carpets, and tableware.

MANUFACTURER'S STATEMENT
Bridgehead is part of an international movement of alternative trading organizations, bypassing large commercial enterprises and going directly to small-scale artisans' and farmers' groups that are set up for the benefit of all. The guiding principles are:

- To pay fair prices for the product of their labor within the context of their economy.
- To seek out producer/partners who have social as well as economic objectives (i.e., literacy classes, health care, and awareness raising programs) that will work toward reducing inequalities in their communities and internationally.
- To honor the orders placed even if producer/partners have difficulty meeting the schedule. Many producers are working under conditions of political violence and are unable to get orders completed or shipped out of the country on time.
- To pay up to 50% of the order in advance, as the producers don't have access to credit to buy raw materials.
- To return part of the profits to the Third World partners as a Development Dividend (5% of the landed cost of the products, equaling $80,000 to be divided between 38 partner groups).
- To consider environmental impact and worker health and safety when choosing products. Part of the Development Dividend is set aside for reforestation where Bridgehead buys wood products (although they do not believe that woodcarvers are responsible for the loss of the world's forests, when compared to massive logging and agriculture operations).
- To seek products which reflect the cultures of the producer/partners. Providing a market for traditional skills and art forms helps preserve them in the face of industrialization.
- To operate a model of just economic relations at home, as well as abroad, by offering products by Native Canadian cooperatives, and employing disabled adults to help with packaging and labeling.

PRODUCTION

SUSTAINABLY ACQUIRED, OR RENEWABLE RESOURCE

IN-PLANT ENERGY EFFICIENCY & RECYCLING

PACKING / SHIPPING

MINIMUM, RECYCLED, RECYCLABLE PACKAGING

INSTALLATION / USE

LOW TOXIC EMISSIONS IN USE

RESOURCE RECOVERY

OTHER

FAIR BUSINESS PRACTICES

RESEARCH & EDUCATION PROGRAMS

PRODUCTION

SUSTAINABLY
ACQUIRED, OR
RENEWABLE
RESOURCE

PACKING / SHIPPING

INSTALLATION / USE

LOW TOXIC
EMISSIONS
IN USE

DURABLE

SIMPLE,
NONTOXIC
MAINTENANCE

RESOURCE RECOVERY

OTHER

PRODUCT
Pure fiber bedding (mail-order)

MANUFACTURER
The Cotton Place
P. O. Box 59721
Dallas, TX 75229
Tel (214) 243-4149, (800) 451-8866

DESCRIPTION
NATURALGUARD™ CLOTH MATTRESS AND PILLOW COVERS: Made of finely woven, pure finish cotton barrier cloth, in white only. Fully washable and dryable at medium temperature. Mattress cover available in twin, long twin, full, queen, king and crib. Pillow covers available in standard, queen, king, and youth. Custom sizes on special order.

DUAL COVER BLANKET: 100% pure finish cotton with tufted terry construction, made in Israel. Provides the same warmth as thermal blanket. Long lasting, preshrunk. Size: 74″ × 94″. Available in white or natural, and 6 colors.

BATH BLANKET: Soft unbleached flannel blanket, containing no permanent chemical finish. For use as sheets or light blanket. Size: 70″ × 90″.

SHEETS AND PILLOW CASES: Wamsutta 100% cotton percale sheets, with no chemical processes, but a natural finish. White only. Fitted bottom and flat sheets available in twin, long twin, full, queen, and king sizes. Pillow cases, sold in pairs, available in standard and king sizes.

CRIB SHEET: White cotton knit sheet. Size: 28″ × 52″.

MUSLIN SHEETS AND PILLOW CASES: 100% cotton muslin, 140 thread count, in white only. Standard pillow case, flat sheet (72″ × 120″).

MATTRESS PADS: Woven pads with stitched edges, in white only. 18 oz., double-napped, all cotton with fleecy surface. Machine wash and dry. Available in twin, full, queen, and king sizes.

MANUFACTURER'S STATEMENT
The Cotton Place specializes in products suitable for persons with chemical sensitivities. Naturalguard™ barrier cloth is custom manufactured and contains no permanent chemical finish. Made from pima cotton in an Oxford weave, with 280 thread count.

PRODUCT
Pure wool duvet (mail-order)

MANUFACTURER
Island Sheppard International Inc.
Hillsborough, Mount Stewart
PEI C0A 1T0
Tel (800) 565-0264 Canada, (800) 565-9070 U.S.
Fax (902) 676-2806

DESCRIPTION
FOUR SEASONS DUVETS: Two-unit ensemble that is zipped together for colder climate and used individually for warmer weather. Wool is enclosed in 100% cotton ticking. One unit weighs 16 oz. per yard (400 gr. per square meter)—for use in spring and fall. The other weighs 10 oz. per yard (250 gr. per square meter)—for summer use. Available in twin, double, queen, and king sizes. 10-year warranty.

UNDERBLANKET: Available in twin, double, queen, and king sizes.

MANUFACTURER'S STATEMENT
The duvet is strong but light, cool in warm weather, warm in cool weather. Can soothe arthritis and rheumatoid aches and pains due to environmental humidity. Nonallergenic for those with allergies to down and feathers.

PRODUCTION

SUSTAINABLY ACQUIRED, OR RENEWABLE RESOURCE

PACKING / SHIPPING

INSTALLATION / USE

LOW TOXIC EMISSIONS IN USE

DURABLE

SIMPLE, NONTOXIC MAINTENANCE

RESOURCE RECOVERY

OTHER

PRODUCTION

SUSTAINABLY
ACQUIRED, OR
RENEWABLE
RESOURCE

PACKING / SHIPPING

INSTALLATION / USE

LOW TOXIC
EMISSIONS
IN USE

DURABLE

SIMPLE,
NONTOXIC
MAINTENANCE

RESOURCE RECOVERY

OTHER

PRODUCT
Pure cotton mattresses and bedding (mail-order)

MANUFACTURER
Janice Corporation
198 Route 46
Budd Lake, NJ 07828

DESCRIPTION
MATTRESS: 100% cotton custom-made mattresses and boxsprings (orthopedic). Plush cotton felt around heavy coils, with no synthetics. Available in *firm* or *superfirm*.

Available in twin, double, queen, king, crib mattress, and hospital mattress sizes. 10-year warranty. With physician's prescription, products are custom-made without fire retardants.

QUILTS: 100% cotton quilts and 100% lambswool quilts, with prewashed combed cotton ticking or cotton barrier cloth. Available in twin/double, double/queen, and queen/king sizes.

MATTRESS PADS: Wool: Quilted fleeced wool on sleeping side, pure wool batting and 100% cotton knit on underside. 1½" thick. All natural, no dyes, no fire retardant.

Cotton: Quilted cotton with 100% cotton top, filling, and bottom, without additives.

SHEETS: Wamsutta 100% cotton percale sheets, without permanent press finish. White only. Available in fitted and flat twin, double, queen, and king sizes.

Cotton knit sheets: 2-way stretch fabric without elastic, 100% combed knit cotton. White only. Available in crib, twin, and double sizes.

BLANKETS: *Top of the Line*: 100% cotton thermal blanket. Colors: blue, rose, cream, white. Available in twin, full, queen, and king sizes.

Reverie Economy: 100% cotton thermal blanket. Colors: tan only. Available in twin, full, queen, and king sizes.

Ether Blanket: institutional quality 100% unbleached cotton with stitched edge. No dye, bleach, or synthetics.

Hudson's Bay Point wool blanket: 100% wool, exclusive weave, self-binding, no fire retardants, no additives. Color: natural background with green, yellow and black stripe. Available in twin, full, queen, and king sizes.

BEDSPREADS: *Stonebridge*: all-cotton, classic herringbone pattern with decorative fringe edge. Machine wash and dry. Color: tan only. Available in twin, full, queen, and king sizes.

Plymouth Historic: 100% cotton with knotted string fringe. Color: bone only. Available in twin, full, queen, and king sizes.

MANUFACTURER'S STATEMENT
100% cotton bedding can provide comfort and allergy relief for those with chemical sensitivities. All activities of production (from initial purchase of raw cotton to last finishing stitch) are monitored and supervised to ensure that materials are free of pollutants and additives. No oils and chemicals are used. Purchased cotton ticking is washed and rinsed twice to remove impurities.

Pure wool & cotton bedding (mail-order)

MANUFACTURER

Pure Podunk, Inc.
RR#1 Box 69
Thetford Center, VT 05075
Tel (802) 333-4505

DESCRIPTION

TOP MATTRESSES: Pure wool in tough, untreated, prewashed 100% cotton fabric case, sewn with cotton & linen thread. 3" thick. Available in baby, twin, double, queen, and king sizes.

INNER-SPRING MATTRESSES & BOXSPRINGS; FUTON: 100% organic cotton cover filled with organic wool batting. Available in twin, double, queen, and king sizes. Baby (futon only).

SHEETS, PILLOW CASES, COMFORTER COVERS: 100% cotton, unbleached, un-dyed, formaldehyde-free, natural beige color, 200 thread count. Available in twin, double, queen, and king sizes. Baby (pillow cases only).

COMFORTERS: Wool and cotton comforters, available in 3 weights: *Summer weight*, *Regular weight*, and *Delux*.

PILLOWS: Wool-filled pillows in soft and firm. Available in baby (boudoir), standard, continental, and king sizes.

MEDITATION CUSHIONS: Same construction as mattresses, available in 2 styles: *Combination*, consisting of mat and cushions, and *Sole cushion*, with carrying handle.

MANUFACTURER'S STATEMENT

Toxic chemicals have not been used, either on the farm, or in production. Pure Podunk depends on small New England and New York farms for supply of chemical-free wool. The wool works are solar powered, without use of oils, flame retardants, or other additives. Materials are organic and chemical free.

A donation of 1% of profit is given to the Working Land Fund, which supports existing and develops future rural communities. Pure Podunk Financial Aid Fund accepts contributions to enable people with chemical sensitivities to purchase safer products.

Pure Podunk has a passionate concern for personal and environmental health, and a desire to see vital, sustainable rural communities. Pure Podunk does not pollute the environment, exploit labor, or mistreat animals. Pure Podunk is committed to the revival of agragarian economy and culture, by supporting local nonchemical farms, sustainable yield logging operations and operating as a rural cottage industry.

Pure Podunk donates 7% of its profits to charitable organizations concerned with these issues.

PRODUCTION

SUSTAINABLY ACQUIRED, OR RENEWABLE RESOURCE

IN-PLANT ENERGY EFFICIENCY & RECYCLING

LOW EMISSIONS PLANT

PACKING / SHIPPING

INSTALLATION / USE

LOW TOXIC EMISSIONS IN USE

DURABLE

RESOURCE RECOVERY

OTHER

FAIR BUSINESS PRACTICES

RESEARCH & EDUCATION PROGRAMS

METRIC/IMPERIAL
CONVERSION TABLE

WEIGHTS

1 gram	=	0.032 ounce
1 ounce	=	28.35 grams
1 kilogram	=	2.21 pounds
1 pound	=	0.45 kilogram

LENGTHS

1 centimeter	=	0.39 inch
1 inch	=	2.54 centimeters
1 foot	=	30.48 centimeters
1 meter	=	1.09 yards
1 yard	=	0.91 meter
1 centimeter^2	=	0.16 inches^2
1 inch^2	=	6.45 centimeters^2
1 foot^2	=	0.09 meters^2
1 meter^2	=	1.2 yards^2
1 yard^2	=	0.84 meters^2
1 centimeter^3	=	0.061 inches^3
1 inch^3	=	16.39 centimeters^3
1 foot^3	=	283.17 centimeters^3
1 meter^3	=	1.31 yards^3
1 yard^3	=	0.77 meters^3

GLOSSARY

Acetate: A chemical structure, usually referring to cellulose acetate fiber or film, or vinyl acetate used in adhesives. *See also* Cellulosic; Polyvinyl acetate.

Acetone: A very volatile and toxic solvent used in lacquers, inks, and adhesives, and for cleaning equipment used for fiberglass resin application.

Acid rain: Rain or snow which is abnormally acidic due to atmospheric pollution. Acid rain destroys lakes, kills trees, and damages buildings. The major pollution sources causing acid rain are coal-burning electrical generators, automobiles, oil and gas furnaces, metal smelters, and other industries.

Acrylic impregnated flooring: Prefinished sheet flooring system that has had liquid acrylic forced under pressure into its porous structure. The acrylic hardens, forming an extremely abrasion-resistant finish, throughout the full thickness. Dyes and fire retardants may be added.

Acrylic polymers: A family of plastic materials used for rigid plastic sheets (Plexiglas), liquid coatings (floor wax and sealers), paints, and many other products. Acrylics are made from acrylic acids, methacrylate, or acrylonitrile, all derived from petroleum. Acrylics are relatively stable and low in toxicity.

Additive: Any material which is mixed with another material to alter its properties. Additives may be used for many purposes, such as preventing hardening or softening, insect attack, or deterioration from light.

Admixtures: An additive, especially a minor ingredient, mixed with concrete to impart particular properties, such as color, decreased drying time, or improved workability.

Aggregate: A granular material used in a mixture, such as crushed rock and sand in concrete.

Aliphatic hydrocarbons: A large family of chemicals based on hydrogen and carbon atoms connected in straight or branched chains. Aliphatics include many paraffins and oils, petroleum derivatives, and bases for many plastics. *See also* Aromatic hydrocarbons.

Allergen: Any substance capable of producing an allergic response, especially in sensitive persons. Some common allergens are proteins contained in pollens, grains, fungi, nuts, and seeds.

Alloyed metal: A mixture of two or more metals, e.g. brass is an alloy of copper and zinc.

American National Standards Institute (ANSI): The U.S. organization responsible for coordinating manufacturing, building, and testing standards.

Ammonia: A very irritating and potentially toxic gas (NH_3), which is the basis of a large number of industrial chemicals and fertilizers. Easily dissolved in water, ammonia is a common component of latex paints and household cleaners. Ammonia is also produced naturally by bacterial action.

Ammonia fumigation: A process using ammonia gas to neutralize formaldehyde emissions from such materials as particle board and adhesives. *See also* Hexamethylene tetramine.

Ammonium sulfide: Usually refers to ammonium bisulfide, a hazardous and irritating chemical used in textile dyeing and treatment.

Aniline dye: Any water-soluble, synthetic, organic dye. It was traditionally derived from the indigo plant.

Antibacterial: A treatment that prohibits the growth

of bacteria both on the surface of a material and within the material itself.

Aromatic hydrocarbons: A large family of chemicals based on hydrogen and carbon atoms which form ring-shaped molecules. Many aromatics evaporate readily and have strong odors (e.g. toluene and xylene). Many are toxic or carcinogenic (e.g. benzene).

Aromatic oils: A fragrant oil, usually a plant extract, with a strong odor, sometimes mixed with a volatile solvent, such as ethyl ether or alcohol. Used as a perfume. Not chemically related to aromatic hydrocarbons above.

Arsenic: A very toxic element used in chemical processing, pesticides, and wood treatments. Arsenic is highly toxic if swallowed or inhaled, and causes skin cancer through skin contact.

Asbestos: A family of mineral fibers found in certain types of rock formations. Asbestos is very fireproof and resistant to chemical attack. It has been used in insulation, plaster, floor tiles, and concrete pipe. The fibers are very hazardous if inhaled or swallowed, and cause lung cancer, chest cancer, asbestosis (a lung disease), and other forms of cancer. Of the several types of asbestos, crocidolite is the most dangerous and chrysotile the least. Asbestos use has been severely restricted in building materials and consumer products since the early 1970s.

Asphalt-treated paper: A facing on batt insulation that provides a barrier to moisture, consisting of paper coated with asphalt (derived from petroleum byproducts).

Axminster: A carpet construction that consists of woven cut pile using 80% wool and 20% nylon, or 50/50 blends of wool and nylon. Manufactured with most of the yarn on the face.

Backed vinyls or laminated vinyls: Consist of a surface wear layer of vinyl and a backing layer, usually of fabric, paper, or plastic foam.

Backing: Plastic coating for upholstery and wall fabric to ensure dimensional stability, or to resist slippage, raveling, and fraying.

Bactericide: Any agent which will destroy bacteria. Bactericides are commonly added to water-based paints, waxes, and other consumer products to prevent spoilage. They are toxic materials which are usually only safe in very low concentrations. *See also* Biocide; Fungicide.

Barite: A naturally occurring mineral, barium sulfate, used in paper and textile coatings, inks, and paints. Low toxicity.

Barrier cloth: A special synthetic or cotton fabric that does not allow dust to penetrate. It has a very high thread count (300 per inch), and is tightly woven.

Batt or blanket insulation: Glass or mineral wool which may be faced with paper, aluminum, or other vapor barrier. For use in walls and ceiling cavities.

BCF: See Bulked continuous filament

Beck dyeing: A method of piece-dyeing small lots of fiber in an open vat, often on a custom basis.

Bentonite: An aluminum silicate clay capable of great expansion when wet. Used in ceramics, paper coatings, waterproofing, foods, and cosmetics. Low toxicity.

Benzene: A carbon and hydrogen compound with a ring-shaped molecule. One of the most common building blocks for synthetic chemicals. Generally derived from petroleum, benzene is highly toxic and carcinogenic. *See also* Aromatic hydrocarbon.

Berber: A carpet construction that consists of a loop pile. Usually made of wool, in natural colors.

Binder: Any material used to hold another together or make it more adhesive. Binders are usually synthetic polymers and are commonly used in paints to form a film when dry.

Biocide or biostat: An additive which will prevent growth of bacteria or fungi. Biocides are used in paints, floor coverings, and sometimes in fabrics. They are toxic materials which are usually only safe in very low concentrations.

Biodegradable: A material which can be decomposed when discarded by the normal action of bacteria and fungi. Typical examples are paper and wood products, natural fibers, and starches.

Biphenyl ingredients: A group of toxic chemical additives based on double benzene rings. Biphenyls (also called diphenyls) are used for dyeing processes and as fungicides.

Bituminous: A substance containing bitumen, which refers to various mixtures of hydrocarbons, such as asphalt, crude petroleum, or tar, often together with their nonmetallic derivatives (dark brown or black in color).

Blanket-wrapped: A form of packaging for shipping furniture, using a returnable blanket. This method leaves no packing materials that require disposal.

Block printing: A hand process, using ink and a block to print on cloth or paper.

Blue Angel (Blauengel): The first environmental listing program, started in Germany in the 1970s, identified by a blue angel symbol. Blue Angel currently lists hundreds of European consumer products, including paints and other building products.

Book-matched veneer: A method of matching veneer to produce symmetric patterns.

Boric acid, boron salt: Naturally occurring compounds containing boron. Commonly used in household cleaners and as fungicides in consumer products. Boron-based fungicides are much safer than their mercury-based or biphenyl counterparts.

Bouclé: A novelty yarn that is looped to produce a pebbly surface.

Breathability: The ability of a finish to allow moisture to escape from behind the film without causing blistering or peeling.

Broadloom: Carpet woven on a loom in continuous rolls, ranging from 6' to 18' wide.

Broad spectrum biostat: A biocide that attacks a wide range of fungus and bacteria types.

Building-related illness: See Sick building syndrome

Bulked continuous filament (BCF): A continuously formed synthetic fiber used in making carpets, which is crimped and fluffed to give more coverage with the same weight, and a fuller feel. It will not shed or pill, and will retain its appearance longer than a staple yarn.

Cadmium: A soft, easily molded heavy metal, used in pigments and heat stabilizers in the vinyl-making process. It is quite toxic and causes permanent kidney and liver damage. Cadmium accumulates in the environment.

Calcination: The heating of minerals to concentrate them, and remove moisture or volatile compounds, e.g. gypsum is calcined to make plaster of Paris.

Calcium carbonate: The mineral base of limestone, marble, seashells, and chalk. Calcium carbonate is heated to form lime (calcium oxide). Very low toxicity.

Calendering: The process of passing hot, dough-like emulsions of vinyl or linseed oil through heated rollers, forming flat sheets. Once cooled, the sheets are trimmed, sanded, and either rolled or cut into tiles for flooring.

Canada Mortgage and Housing Corp. (CMHC): The national housing agency of Canada responsible for housing research and standards, housing assistance programs, and some mortgage insurance and product certification.

Carcinogen: Any naturally occurring or synthetic substance known to increase the risk of cancer.

Carding: The process of disentangling and equalizing the distribution of fibers before spinning it into yarn.

Carnauba: A wax derived from the Brazilian wax palm leaf. Used extensively in paste waxes, furniture pol-

ish, plastics, and food glazes. Very low toxicity.

Carpet tile: Carpet made and laid in small units (2 or 3 feet square), allowing for easier removal and/or replacement.

Case goods: Hard furniture storage components, such as cabinets and files.

Casein (calcium caseinate, sodium caseinate): The protein base of milk. Casein is used as a base and binder in natural paints and adhesives. It is a very low toxicity material but is allergenic to people with milk sensitivies.

Catalyst: A substance used to accelerate a chemical reaction, which is not consumed in the reaction.

Caulking: Compound used to fill joints and cracks in a building's surface, such as around door and window openings.

Cellulose: The natural carbohydrate base for plant fibers, it is the most abundant organic material in the world. Used to make paper and textiles.

Cellulosic: A synthetic material made from cellulose. Acetate, rayon, and viscose are cellulosic fibers.

Cembra pine: Also known as Swiss pine, its resin has inherent pesticide properties.

Cementitious: Any material based on cement or cement-like products, i.e. inorganic, noncombustible, and hard-setting.

CFC: See Chlorofluorocarbons

Chamois: A soft leather prepared from the skin of a small, goat-like animal. Imitation chamois is an oil-tanned, suede-finished leather prepared from the inner layer of split sheepskins. Chamois cloth is a soft fabric made in imitation of chamois leather.

Chemically stable materials: Those which will not readily break down, release chemicals, or change into other (potentially toxic) chemicals with age, heat, or light.

Chemical sensitivity: A loosely defined condition experienced by some people who are affected by very small concentrations of chemicals in air, water, and food, which would not have apparent effects on most people. Symptoms may be similar to minor allergies, though they may also include moderate to severe pains, muscle weakness, dizziness, confusion, and even seizures.

Chemical sensitization: The process of becoming sensitive to chemicals through excessive exposure, e.g. to fungi, wood dusts, or formaldehyde.

Chemical weld: A method of joining two sheets of vinyl flooring by applying a solvent to soften the edges temporarily. *See also* Heat welding.

Chipboard or particle board: A building panel consisting of wood chips and fibers pressed together, using a synthetic resin as a binding agent.

Chlorinated compounds: A large family of synthetic chemicals which are formed by combining chlorine with hydrocarbons. Many, such as DDT, are environmental toxins and very persistent. Some, such as the chlorinated solvents, present serious disposal problems. Many are toxic or carcinogenic, and will accumulate in fat tissues, liver and kidneys. *See also* Chlorofluorocarbons.

Chlorofluorocarbons (CFCs): Synthetic chemicals manufactured from hydrocarbons (usually methane) and chlorine, fluorine, or bromine. Commonly used as refrigerants and solvents, all have some potential to destroy ozone in the upper atmosphere when released. Most are very chemically stable, and some are toxic. Most are being phased out and will be banned by 1997.

Chromate: Compounds containing chromium. Often refers to potassium dichromate, used in dyes and leather tanning. Very toxic and hazardous to the environment.

Clay body: A mixture of clays, or clays and other mineral substances, blended to achieve a specific ceramic purpose.

CMHC: See Canada Mortgage and Housing Corp.

Coalescing solvent: A solvent that causes two or more substances to combine, e.g. a solvent which causes a paint to harden.

Cogeneration: An energy efficiency measure where two power supplies come from the same source to reduce energy losses. A typical cogeneration facility will produce electricity and steam from the same boiler. *See also* Plant energy efficiency.

Colorway: Refers to one color combination from the full coloration line of a carpet or fabric product.

Continuous dyeing: A dyeing method that consists of processing a great bulk of fiber at one time, under pressure. This method ensures consistency of color, and is a closed system, with little waste.

Conversion: A form of recycling in which a waste material is turned into a useful material of substantially lower quality. An example is the use of crushed concrete and bricks as a granular base for roads and sidewalks.

Conversion efficiency: The ratio of raw materials going into a process to product coming out. A sawmill with a high conversion efficiency will produce more lumber and less waste from the same logs as a mill of lower conversion efficiency.

Copolymer: See Polymer

Coproducts: Materials which are incidentally or intentionally produced when making another material. Coproducts which have value or are useful to another industry are beneficial, while those which are not useful become waste.

Corrosive: A substance that causes materials to deteriorate through a chemical reaction.

Cottage industry: A small volume industry, typically operated at home.

Cradle to the grave: The complete lifespan of a material from the extraction of raw materials through manufacturing, installation, use, maintenance, and eventual disposal. *See also* Life cycle costing.

Crimping: A process in which natural and synthetic fibers are set, during the finishing process, into wavy coils for resilience, wrinkle resistance, and natural cohesion.

Curing: The process and time period for a finish to achieve its final state of hardness, color, etc. A chemical reaction is usually involved.

Cut pile: A carpet construction with the entire surface of the pile being sheared, producing a smooth, plush appearance.

Dacron: Trade name for a polyester fiber made from polyethylene terephthalate (PET). *See also* Polyethylene terephthalate.

Dammar: Resins derived from tree saps, used in varnishes, baked enamels, and paper and textile coatings. Low toxicity.

Damp-proofing: A treatment, such as a sealer or asphalt coating, that inhibits the transfer of moisture.

Defoliant: Chemical agent which causes leaves to drop off. Defoliants are used, for example, by many cotton growers to simplify harvesting. Defoliants are toxic substances, and remain in the finished cotton.

Denier: Unit for measuring the fineness of yarn filament. The higher the denier, the finer the yarn.

Diamond-matched veneer: A pattern in which four pieces of veneer from the same material are cut diagonally, then fitted together to form a diamond. A reverse diamond is created when veneers are rotated 90 degrees.

Dimethyl ether: An ether made from wood alcohol, used as a solvent and propellent in spray cans. Moderate toxicity but very flammable.

Direct dyes: A class of dyes used for cellulosics that requires no fixatives.

Direct glue-down: Installation method where the carpet is glued directly to the floor surface, rendering the material unusable upon removal.

Dispersion: A mixture in which solids are suspended in a liquid, e.g. paint, usually in small globules.

Dolomitic: Limestone or chalk materials made up from

carbonate rocks. Very low toxicity. *See also* Calcium carbonate.

Drying oil: An oil, such as linseed or synthetic oil, which contains additives causing it to harden when exposed to air. Usually used in paints.

Dry joint: A joint without mortar between stones or tiles.

Dry-pressed: A method of forming individual tiles by compressing dry clay under pressure, then firing it.

Ecological cost (environmental cost): An assessment of the effects of an action on the environment and all living things. Impacts such as resource depletion, air, water, and solid waste pollution, and disturbance of habitats are typically included. *See also* Environmental audit.

Ecosystems: Networks of biological, geological, and chemical activity which sustain life. The ecosystem of a pond may include aquatic plants, nutrients, soils, bacteria, insects, fish, frogs, and ducks in their mutual dependencies.

Efflorescence: The formation of white crystals on the surface of tile or brick, caused by leaching of mineral salts as the wet material dries. Not hazardous.

Elasticity: The ability of a material to recover its original form or condition.

Elastomer: A rubber-like material which will remain flexible, used to fill cracks, join other materials which may move over time, and where resilience is important. Caulkings are elastomers, as are soft rubber and plastics. *See also* Polymer.

Electrostatic painting: A process in which the substrate is electrostatically charged, causing sprayed paint to cling directly to the surface, reducing waste and worker exposure.

Embodied energy: The energy represented by a building, or part of a building, including all the energy used in extracting and manufacturing materials, shipping, assembling, and finishing them. Embodied energy is one useful measure of the "ecological cost" of a building. *See also* Energy intensity.

Emissions: Releases of gases, liquids, and solids from any process or industry. Liquid emissions are commonly referred to as effluents.

Emissions controls: Any measure which reduces emissions into air, water, or soil. The most effective emissions control involves redesign of the process so less waste is produced at source. Common emissions controls are dust collectors, wastewater treatment plants, and in-plant solid and toxic waste reduction programs. *See also* Stack scrubbers; Secondary treatment; Tertiary treatment.

Emulsion: See Dispersion

Energy intensity: The energy required to make a material, including that used in extracting raw materials and production processes. Transportation and construction energy are usually excluded.

Environmental audit: A study of the environmental impact of a product or process. *See also* Ecological cost.

Environmental Choice/Ecologo: The environmental product listing program of Environment Canada, identified by a logo of three doves intertwined in a maple leaf. The program currently lists some paints, insulations, paper products, energy efficient lighting, and ventilators, as well as other consumer products with environmental merit.

Environmental Protection Agency (EPA): The U.S. national agency responsible for air and water quality standards, industry regulation, waste management, and many other programs.

Environmental restoration: The act of repairing damage to a site caused by human activity, industry, or natural disaster. The ideal of environmental restoration, though rarely achieved, is to leave a site in a state which is as close as possible to its natural condition before it was disturbed. Examples are replanting forests, stabilizing soils, and filling in and replanting mine pits.

Environment Canada: Canada's national agency responsible for air and water quality standards and programs, industry regulations, and statistics. Also operates the Environmental Choice Program, a "green" product listing program.

Epichlorohydrin: A very toxic and carcinogenic substance used in manufacturing epoxies, some synthetic rubbers, and other adhesives.

Epoxy: A class of synthetic resins used for high performance adhesives, paints, and protective coatings. Epoxy adhesives and paints are two-part materials mixed immediately before use. The ingredients are hazardous and should be handled by professionals.

Ergonomics: The study of the "fit" between people and machines or furniture. Ergonomically designed furniture is more comfortable and reduces physical stress while improving productivity.

Ethereal oils: See Aromatic oils

Ethyl alcohol: Grain alcohol, sometimes used as a solvent in paints and waxes, and in dyeing processes. Ethyl alcohol is intoxicating when ingested, but the vapor is relatively harmless.

Ethylene glycol: An alcohol often used as a solvent in water- and oil-based paints, lacquers, and stains. Toxic when ingested or inhaled, ethylene glycol is also the main ingredient of automotive antifreeze.

Exposed aggregate: A mixture of a variety of small stones that is poured or pressed onto a surface with a cement paste. When the concrete is partially hardened the surface is washed or brushed, exposing the aggregate.

Extruding: A method of forming a liquid or solid material into a long thin piece, usually under high pressure. Aluminum is extruded into tubes and channels, and nylon is extruded into fibers.

FDA: See Food and Drug Administration

Feldspar: A natural, silica mineral used in glassmaking and ceramic glazes.

Ferric oxide: Oxide of iron (rust). Used for pigment in paints. Very low toxicity.

Fiberboards: Construction panels made from compressed fibers, including wood, paper, straw, or other plant fibers. Three common types are:
High density fiberboard (HDF), usually made without glues and used for doors, furniture, cabinets, and pegboard;
Low density fiberboard, usually made without glues and used as acoustical insulation and tackboard;
Medium density fiberboard (MDF), usually made with formaldehyde-based glues and used for furniture, shelving, doors, and flooring underlayment.

Fiberglass resin: Uncured polyester resin used for fabricating glass fiber reinforced, and cast plastic products. Irritating and toxic to handle.

Fiberization: The process of reducing a material, such as newspaper or cotton, into a loose fiber.

Filament: A continuous strand of natural or synthetic fiber.

Filler: A material of little or no plasticity which adds bulk to paints or plastics, or helps to promote drying and control shrinkage in clay bodies.

Filling yarn (weft): Yarn carried horizontally on a shuttle through the open space of the vertical warp, in a woven fabric.

Fire retardant or flame retardant: A substance added to a flammable material to reduce its flammability. It is only possible for fire retardants to slow the spread of fire; they do not make flammable materials fireproof. Used in fabrics, carpets, bedding, upholstery, and foamed plastics, some are hazardous phosphates and chlorinated compounds.

Flagstone: Thin slabs of stone used for paving walks, driveways, and patios. Generally fine grained sandstone, bluestone, quartzite, or slate is used.

Flame spread: The speed at which fire will move through a material using standard laboratory testing methods.

Fluoride: A compound of the very active element fluorine. In high concentrations fluorides are toxic to humans and wildlife. Fluoride emissions from factories, such as aluminum smelters, do serious damage to forests, and lake and river systems.

Fluorocarbons: See Chlorofluorocarbons; Hydrogenated chlorofluorocarbons

Flux: A substance which promotes melting or fusion of metals or minerals.

Foamed-in-place: An insulating material containing cements or plastics which is installed wet using foaming equipment.

Food and Drug Administration (FDA): The U.S. national agency responsible for setting safety standards for foods, cosmetics, pharmaceuticals, and food additives. Some "food grade" additives are used in paints and coatings because of their relatively low toxicity.

Formaldehyde: A pungent and irritating gas used in the manufacture of many adhesives and plastics, as a preservative, and in permanent press fabrics. Many people suffer health effects from formaldehyde at very low concentrations. Formaldehyde is now listed as a suspected carcinogen.

Formaldehyde scavengers: Agents added to wood products manufactured with formaldehyde glues to neutralize gases before they escape. Usually sulfite or ammonia compounds, scavengers are sometimes used to treat affected buildings directly by fumigation, or as a coating for air filters. *See also* Ammonia fumigation; Sulfites.

Fossil fuels: Oil, gas, or coal fuels derived from ancient vegetation. Fossil fuels were formed several hundred million years ago when seas and swamps covered most of the continents. *See also* Nonrenewable energy.

Freon: A trade name for CFC refrigerants. *See also* Chlorofluorocarbons.

Fungi (molds, mildew, mushrooms): Plant-like organisms which do not require light for growth and survival because they do not produce chlorophyll. Many fungi are safe and edible, such as some species of mushrooms, while others produce very toxic and allergenic substances.

Fungicide: Any substance added to inhibit the growth of fungus and consequent spoilage of a material. Paints, stucco, floor coverings, treated wood, and outdoor fabrics are commonly treated with fungicides. Many are hazardous metal or chlorine compounds. Safer fungicides are usually boron or sulfate based. *See also* Biocide.

Gauge or pitch (carpeting): The number of rows across a carpet, expressed as a quantity of stitches per inch. *See also* Stitch count.

Gauge (flooring): The overall or nominal thickness of a flooring material. Gauge affects the cost of the material as well as its suitability for certain applications.

Generally regarded as safe (GRAS): Materials which do not require toxicity testing due to their long history of use without apparent ill effects. Shellac and linseed oil, for example, have been in use for centuries, as have many pigments from plants.

Glass fiber: Glass which has been extruded (stretched) while molten to make very fine fibers for insulation, and nonflammable fabrics or reinforced plastics. Glass fibers are irritating to the skin and dangerous if inhaled, though not as hazardous as asbestos.

Glaze: A glassy surface treatment and/or embellishment fused onto clay during its final firing. There is a wide range of color, opacity, and reflectance available.

Global warming: The predicted increase in average temperatures over the earth due to increased carbon dioxide, nitrogen oxides, methane, CFCs, and other gases released into the atmosphere by human activity. Altered weather patterns, rising sea levels, and changing agricultural potential are likely outcomes.

Glycols: A family of alcohols used as solvents in many paints, coatings, etc. Some glycols are very safe while others are toxic. *See also* Ethylene glycol; Propylene glycol.

GRAS: See Generally regarded as safe

Green Cross: An American environmental listing program, identified by a green cross symbol. Listed products have environmental merit in terms of energy efficiency, low toxicity, and recycled content.

Greenhouse gas: See Global warming

Grout: A cementitious material used to fill the joints between tiles. May contain latex or epoxy additives for greater durability.

Hammermill refiner: A machine using mechanical hammers for crushing stone, clay, ore, etc.

Hardwood: Usually wood from a deciduous tree (one that loses its leaves in winter). Hundreds of hardwoods are known from both temperate and tropical regions, and most species are slow growing. They are generally used for furniture and flooring.

HCFC (or HFC): See Hydrogenated chlorofluorocarbons

HDF: See Fiberboards

Heat welding: A process of joining two sheets of resilient flooring by heating and inserting a color-matched welding thread of polyvinyl chloride along the length of the seam. Once cooled, the welding thread is trimmed flush with the floor, creating a waterproof seam.

Heavy metals: The series of metals including mercury, lead, cadmium, thallium, cobalt, nickel, and aluminum. Most are very toxic and persistent in the environment. Mercury and lead accumulate in tissue and cause nervous system damage, cobalt and thallium are extremely toxic, cadmium is toxic to the kidneys and liver, aluminum is associated with Alzheimer's disease, and nickel is carcinogenic.

Hessian: A burlap fabric made from jute or hemp.

Hexamethylene tetramine (methenamine, aminoform, or hexamine): A stable, crystalline substance formed by the action of ammonia on formaldehyde. It is the residue of formaldehyde neutralization which remains in wood products and buildings fumigated with ammonia. It is a skin irritant but is relatively low in toxicity.

Hexane (n-hexane): A solvent derived from petroleum, used in adhesives and paints. Hexane is moderately hazardous in low concentrations, causing symptoms of nerve toxicity, such as numbness, trembling, or disorientation.

High build finish: A multiprocess finish that results in a thick coating intended to protect the material.

Homogeneous: A mixture in which all the components are evenly dispersed. A homogeneous resilient floor has color dispersed throughout its thickness.

Honed finish: Surface finish with dull sheen, without reflections.

Housing and Urban Development (HUD): The U.S. national housing agency.

Hydrocarbons: A vast group of naturally occurring and synthetic compounds based on hydrogen and carbon atoms only. The primary source of the world's hydrocarbons is fossil fuels. Hydrocarbons are used as fuels and as feedstocks for most organic chemical synthesis. *See also* Aliphatic hydrocarbons; Aromatic hydrocarbons.

Hydrochlorofluorocarbons: See Hydrogenated chlorofluorocarbons

Hydrogenated chlorofluorocarbons (HCFC or HFC): Hydrogenated chlorofluorocarbons (or hydrogenated fluorocarbons) are substitute refrigerants and solvents which do not have as much potential to destroy atmospheric ozone if released into the atmosphere as do CFCs. Most are less efficient as refrigerants than CFCs and some are quite toxic. *See also* Chlorofluorocarbons.

Hydrogen sulfide (H_2S): A poisonous and odorous gas found in natural gas deposits, and produced by bacteria when decomposing waste without oxygen present. Gypsum building products present a disposal problem in wet climates because hydrogen sulfide is produced by bacteria when gypsum is buried.

Hydropulp: A mechanical method of breaking down wood fiber into pulp using water pressure instead of caustic chemicals. A very low emission pulping method.

Igneous rock: The oldest type of stone, cooled from melted rock. Included in this category are the hardest stones—granite, porphyry, gabbro and serpentine.

Industrial waste: Materials which are inadvertently produced by manufacturing processes and usually have no commercial value. Waste may be simply a nuisance and a minor disposal problem, or it may be hazardous, or highly toxic. The ideal solution is for waste from one industry to be used as raw material in another industry.

Industry standards: Standards imposed on an industry, either by the industry or by a regulatory agency, to control product quality, manufacturing processes, or plant emissions.

Inorganic compound: Any compound which does not contain carbon atoms in its structure. Minerals, metals, ceramics, and water are examples of inorganic compounds. Most tend to be very stable and persistent because they oxidize slowly or not at all. *See also* Organic compound.

Isocyanate: Similar to Isocyanurate, below.

Isocyanurate: A family of resins, usually called polyurethanes, which are used for insulating and upholstery foams, paints and varnishes. Because isocyanurates contain a cyanide group in their chemical structure, most will release deadly cyanide gas if exposed to a fire.

Isoparaffinic hydrocarbons: A type of highly purified petroleum solvent sold as "odorless paint thinner." Often used in solvent-based, low toxicity paints, it is one of the safest solvents because the odorous portions which have been removed are also the most hazardous.

Jacquard weave: A fabric woven on a loom equipped with a Jacquard device. Named after its inventor, these looms are capable of producing elaborate textiles such as brocades, matelassés, tapestries, and damasks.

Joint compound: A wet filler material used to join materials of the same type, to create a uniform surface, e.g. gypsum filler.

Kerf: Saw-cut on the underside of lumber to permit bending.

Ketones: A chemical structure common to many solvents, such as acetone and methyl ethyl ketone.

Kiln-dried: A method of drying wood in an oven after sawing that results in 10% or less moisture content. This makes the wood more dimensionally stable and better able to resist decay.

Kraft: A papermaking process using softwoods and sulfites. The largest volume paper production process in the U.S. and Canada. Kraft is usually light brown in color because not all of the lignin has been removed by digesting and bleaching.

Lacquer: A glossy liquid finish for woods or metals, traditionally prepared from plant resins. Nitrocellulose lacquer, made from wood or cotton fiber treated with acid and dissolved in butyl acetate (lacquer thinner) is a more typical formulation today. Lacquer also refers to many types of hard, high gloss industrial finishes, such as acrylic auto finishes.

Laminate: A thin layer of material (veneer) bonded to another surface. Wood and plastics are both commonly laminated.

Landfill: Solid waste disposal sites. They may be minimally regulated and contain hazardous materials, or highly controlled to prevent ground water pollution and other effects.

Lanolin: The natural oil that wool contains. It inhibits soiling and gives a natural luster and softer feel to the material.

Latex: A naturally occurring, sticky resin from rubber tree sap used for rubber products, carpet backings, and paints. Latex is a broad term which also can apply to synthetic rubbers, usually styrene-butadiene. *See also* Styrene-butadiene rubber.

Life cycle costing: A method of calculating the total cost of a material or component, including its maintenance and replacement costs over the life of a building. Recently some life cycle costing methods have been been expanded to include ecological costs.

Lignin: A naturally occurring polymer in wood which keeps the fibers bound together. Most lignin is removed during papermaking and is burned as fuel. Pulped wood which has not had the lignin removed can be pressed into high density fiberboard without adhesives because the lignin will bind the fibers.

Linseed oil: Oil from the seed of the flax plant. Used in paints, varnishes, linoleum, and synthetic resins.

Linseed oil putty: A mixture of linseed oil and finely ground calcium carbonate (chalk) used for woodwork and glazing.

Lithopone: A white pigment made from zinc sulfide, zinc oxide, and barium sulfate. Low toxicity.

Loosefill insulation: Vermiculite, perlite, glass or mineral wool, shredded wood and paper. For use in wall cavities and insulated ceilings.

Magnesium oxide: A mineral product used extensively in ceramics, papermaking, cosmetics, and pharmaceuticals (magnesia). Low toxicity, but the dust is hazardous.

Mastic: Originally a resin from the pistachio tree, used as a crack filler and caulking. Now refers to many synthetic caulkings and adhesives used for floors and tile laying. *See also* Elastomer.

Material Safety Data Sheet: A legal requirement for all potentially hazardous products, the data sheet indicates the risks from using and disposing of the product and recommends safe practices. The sheet may also indicate the chemical contents of the product.

MDF: See Fiberboards

MEK: See Methyl ethyl ketone

Melamine: A polymer used for plastics and paints made from formaldehyde, ammonia, and urea. Similar to urea formaldehyde resin.

Mercerizing: A treatment for cotton in which it is passed through a caustic bath and washed, under tension, to preshrink and strengthen it.

Metamorphic rock: Formed from igneous or sedimentary deposits which have been subjected to intense stress, heat, or chemical effect. Slates and marble are metamorphic.

Methyl alcohol: Also called wood alcohol or methanol, it can be made by heating wood or peat under pressure, but is usually made from natural gas (methane). It is far more toxic than ethyl alcohol and is used in shellacs, waxes, and paints.

Methyl cellulose: A product of wood pulp used as a thickener, adhesive, and food additive. Very low toxicity.

Methylene chloride: A solvent, paint remover, and blowing agent for foams. Very toxic and a known carcinogen.

Methyl ethyl ketone (MEK): A common solvent used in laquers, paint removers, and adhesives. Moderately toxic. Methyl ethyl ketone peroxide is a related compound, used as a hardener for fiberglass resin. It is very toxic.

Mica: A naturally occurring silica mineral used as a filler in paints, gypsum fillers, and as electrical insulation. Low toxicity, but the dust is hazardous.

Microbial: See Microorganism

Microorganism: Any microscopic living thing, such as bacteria and some fungi.

Mil: A unit of measurement, one thousandth of an inch.

Monocottura: A tile that has been fired only once. Bicottura tiles are fired twice, when effects or patterns on the glaze are desired.

Mortar: A cement-based mixture used to lay stone or ceramic tiles, or for use as a grout for these materials.

Mosaic tile: Ceramic tiles that are less than 6 square inches. The thickness ranges from ¼" to ⅜".

Mothproofing: A treatment applied to fibers to resist damage by moths. Mothproofings are typically skin irritants and may cause adverse reactions on contact. Wool is the fiber most prone to moth attack.

Muslin: A plain-woven, uncombed cotton fabric, ranging from sheet-fine to coarse textured.

Mutagenic: A substance which is known to alter the genetic material (DNA) of cells, potentially causing changes in tissue growth leading to cancers.

Mylar: Trademark for a polyester sheet used for decorative and industrial purposes. It is very tough, heat resistant, and chemically stable.

Natural: A substance or material which is taken from nature as directly as possible with minimal intervention of processing or chemical synthesis. Though not all natural materials are safe, their properties are well known through traditional uses.

Neoprene: Trade name for polychloroprene, a synthetic rubber used to manufacture caulking, rubber gaskets, and waterproof membranes.

Nitrocellusose lacquer: See Lacquer

Nominal: The named thickness of a material, though not always the actual thickness, e.g. 2" nominal lumber is actually 1½".

Nonrenewable energy: Energy derived from fuels (oil, gas, and coal) produced by geologic processes which take millions of years. *See also* Fossil fuels.

Nonrenewable material: Materials which are not replaced by photosynthesis, i.e. minerals and metals.

Nylon: Any of a family of polyamide resins used for textile fiber, rope, and molded plastics. The most common type in carpet manufacture is Nylon 66, though Nylon 6 is also used. Some nylons are technically recyclable, though very little recycling is currently done.

Occupational Safety and Health Administration (OSHA): The U.S. agency responsible for setting health standards for the workplace.

Offgassing (or outgassing): The release of gases or va—pors from solid materials. It is a form of evaporation, or a slow chemical change, which will produce indoor air pollution for prolonged periods after installation of a material.

Open office furniture system: A system using furniture screens (instead of drywall partitions) to provide visual and acoustical privacy. Usually modular, the parts are interconnecting and the system usually offers electrical wire management features. Less square footage is needed for an open office system workstation than a traditional office.

Organically grown: Grown with minimal use of synthetic fertilizers or pesticides. Various state and industry definitions are used to determine which products can be sold as organically grown.

Organic compound: Any chemical compound based on the carbon atom. Organic compounds are the basis of all living things; they are also the foundation of modern polymer chemistry. There are several million known and their characteristics vary widely.

Organochlorine (organobromine, or organofluorine): Organic compounds formed with chlorine, bromine, or fluorine in their structure. Organochlorines are usually either very toxic and persistent in nature (dioxins, PCBs, DDT) or very chemically stable (PVC, Teflon).

Oriented strand board (OSB): A manufactured wood product, mainly used in construction, composed of strands of wood laid in the same direction and glued together. This process produces a very high strength product from low grade waste material from the wood milling industry.

OSHA: See Occupational Safety and Health Administration

Outgassing: See Offgassing

Oxford weave: A textile made in a basket weave pattern.

Ozone depletion: The loss of atmospheric ozone, the very high altitude layer which protects the earth from destructive ultraviolet radiation. CFCs used as refrigerants and solvents are the main substances involved. *See also* Chlorofluorocarbons.

Paraffin wax: A low toxicity petroleum wax used in some finishes and for making candles. Approved as a food additive.

Parquet: A small wooden tile made from interlocked strips of hardwood, and used as a floor covering. Parquet is usually glued down and varnished.

Particulates: Particles of dust, mold, mildew, etc. small enough to become suspended in air. Very small particulates (less than .005 millimeter) can be inhaled deep into the lungs. Particulates containing plant or animal proteins are allergenic, while those containing mineral fiber (silica, asbestos) cause lung disease or cancer.

Parts per million: A unit of measurement for very small concentrations of gases or liquids. One part per million indicates a concentration of one million to one by volume.

Pattern repeat: A total design unit, repeated over the width and length of the material.

Penetrating/impregnating sealer: A finish that will penetrate the porous structure of wood or tile, and protect not only the surface, but the entire upper layer of the material.

Perchloroethylene (PERC): A dry cleaning solvent and degreasing solvent. Hazardous to handle and dispose of. Vapor is toxic.

Perlite: A volcanic glass which is expanded and used as a plaster additive and fire resistant insulation.

Pesticide: A chemical agent that kills or inhibits infestation of pests, especially insects.

Petrochemicals: Any chemicals synthesized from petroleum. All are hydrocarbons. *See also* Hydrocarbons.

PH: An index of the acidity or alkalinity of a substance. A pH of 1 is highly acidic, 7 is neutral, and 13 is highly alkaline.

Phenol formaldehyde: An adhesive resin used for exterior plywood and other wood products. Dark brown in color and low in formaldehyde emissions.

Phenols (phenolic odor): A hydrocarbon with a ring structure, derived from oil or coal. Used to synthesize a large number of resins, glues, and pharmaceuticals, and as a germicide. Very toxic.

Phenyl mercuric acetate: An organic compound of mercury previously used as a fungicide in paints, but now banned for most interior paints. A highly toxic material which does nerve damage.

Phthalates: A large family of hydrocarbon compounds added to plastics to keep them soft. Also used as solvents and catalysts. *See also* Plasticizers.

Piece dyed: A dyeing method that consists of the yarn being dyed after tufting.

Pigment: Minute solid coloring particles for paints, stains, plastics, etc. Historically derived from clays and other earthen minerals or plants, modern pigments are mostly synthetic.

Pile height: The height of the face yarn or length of the tuft, measured from the top of the pile to the backing.

Pima cotton: An extra-long staple fiber with a silky sheen and exceptional strength and firmness, developed in the southwestern U.S. by selective crossbreeding of Egyptian and North American cottons.

Plain-sawn: Boards cut perpendicularly to the annual

rings, not toward the center of the log, producing wide grain, with V-shapes that tend to wear unevenly. *See also* Quarter-sawn.

Plain weave: Interlacing warp and filling yarns to form a pliable fabric, characterized by an over-and-under pattern.

Plant energy efficiency measures: Redesign of processes or equipment, or the addition of new equipment, which reduces energy requirements or recovers waste energy. Efficient motors and heat recovery units are common examples.

Plasticizer: A chemical added to a plastic or rubber to keep it soft and flexible, particularly common in vinyl upholstery and flexible floor coverings. The plasticizers offgas slowly.

Plywood: An odd number of veneers glued together with the grain of adjacent sheets at right angles to each other.

Polyamide resin: The family of synthetic resins called nylons. *See also* Nylon.

Polyester: Any long chain polymers made from esters of alcohols. Fibers, solid and sheet plastics of many kinds are classified as polyester.

Polyethylene: A chemically simple, semitransparent plastic. Used widely as vapor barrier sheet over insulation, for packaging film, and containers. There are both high density (HDPE) and low density (LDPE) varieties. It is a low toxicity material and produces low risk vapors when it is burned.

Polyethylene terephthalate (PET): A polyester resin used to produce polyester fiber and sheet plastics, e.g. recyclable soft drink bottles.

Polymer or copolymer: Any of a large number of natural or synthetic, organic or inorganic compounds composed of very large molecules that are made up of many light, simple molecules chemically linked together in long chains. Cellulose and proteins are naturally occurring polymers, while plastics are synthetic.

Polyolefin: A class of common, synthetic plastics including polyethylene and polypropylene. Polyolefin can be spun into fibers to produce a tough exterior weather barrier fabric for construction.

Polypropylene: A polyolefin with good flexibility used for carpet yarn, rope, artificial turf, packaging, and primary carpet backing. It has a very smooth surface texture and tends to produce abrasive textiles.

Polystyrene: A plastic used in its foamed form as building insulation or in its hard form for tough, molded plastic products.

Polyurethane: See Isocyanurate

Polyvinyl acetate (PVA): A plastic usually used in water-based emulsion glues, such as white glue, and in

waxes and flooring adhesives. Relatively low toxicity.

Polyvinyl chloride (PVC): A chlorinated vinyl plastic which is very durable and chemically stable unless plasticized to keep it soft. The basis of most flexible flooring, plastic upholstery, and plastic siding. Produced in a closed process using vinyl chloride, a very hazardous material. Vapors are hazardous when it is burned.

Porcelain: A glass-bodied, impervious, dense and homogeneous tile with no surface porosity. It is unaffected by chemical and physical agents such as fire or frost.

Portland cement: A kind of cement made by burning limestone and clay in a kiln. It is the base for most grouts used with ceramic mosaics, quarry, and paver tiles.

Post-commercial waste (industrial waste): Waste from industry processes or materials discarded from industrial plant renovations or maintenance.

Post-consumer waste: Goods or materials discarded by consumers after use.

Pressure-sensitive adhesives: Dry adhesives that adhere with pressure and remain flexible.

Primary backing: Applies to tufted carpets only. A mesh (usually polypropylene) to which the tufts of yarn are anchored.

Primary design consultant: Typically an architect or interior designer. The person responsible for an overall project, the hiring of subconsultants, and communication with the client.

Primary production: Generally involves only the extraction of raw materials.

Primary treatment: Basic wastewater treatment which only removes the bulk of solids but does not digest nutrients or render water safe for discharge. *See also* Secondary treatment; Tertiary treatment.

Primer: First coat applied to a substrate, serving as a sealer and providing a good base for finishing coats.

Product life cycle: See Cradle to the grave.

Propylene glycol: An oily alcohol used in paints, waxes, and sealers. Unlike ethylene glycol it is low toxicity, and approved as a food additive.

PVA: See Polyvinyl acetate

PVC: See Polyvinyl chloride

Quarry: An area with a deposit of one type of stone, which is drilled, blasted, or split from the rock face, and cut into manageable blocks for transporting to facilities for conversion to building materials.

Quarry tile: A strong, hard-bodied tile made from graded shale and fine clays, with the color throughout the body.

Quarter-sawn: Logs are quartered, and boards are sawn toward the center, producing a grain pattern of narrow, parallel, longitudinal strips. Although there is less tendency to shrink or warp, there is more waste of the raw material. *See also* Plain-sawn.

Radon: A colorless, odorless, radioactive gas occurring naturally in soil and rock. Radon exposure increases risk of lung cancer. Most radon enters homes through basements and floor drains.

Raschel knitted: A carpet construction in which the carpet is knitted on a Raschel warp-knitting machine.

Recycling: True recycling is the conversion of a waste material back into its original material. A variant is conversion into another material.

Regeneration: Restoration of logged forest lands and mined sites. Drainage, soil replacement, replanting, and fertilization are usually involved.

Remanufacturing: The reconditioning and repairing of furniture, whereby materials may be completely or partially stripped and replaced, for reuse.

Render: The application of stucco or cement mortar to the face of a wall to give a continuous surface finish.

Renewable: A material produced directly by photosynthesis, i.e. one that can be replaced within a few decades or less.

Rep: A plain-woven fabric characterized by raised, rounded ribs, running from selvage to selvage, producing a linear texture in the fabric.

Resilient: Ability to "bounce back," or return to the original form, after subjecting a material to static (stationary) loads, dynamic (moving) loads, or sudden impact. Rubber, vinyl, and linoleum floor coverings are resilient.

Resin: A sticky substance that flows from certain plants and trees, especially pine and fir. It is used in medicine and varnish. Artificial resins, used in the manufacture of plastics and synthetic finishes, are usually petroleum based polymers.

Reuse: The recovery of a material to be used again for a similar application without reprocessing.

Rigid insulation: Insulation materials, such as foamed plastic, wood, cork, glass or mineral fibers pressed into a standard-sized board for easy handling. Used as a surface insulation.

Rock wool or mineral wool: Insulating material spun from heated slag (waste), from metal smelting. Similar to glass fiber.

Roller printing: For less expensive wall coverings, this process involves inks on a metal roller transfered to paper. *See also* Block printing.

Rubbed finish: Finish with a flat surface and occasional slight "tralls" or scratches.

R-value: Thermal resistivity, that is, the inverse of the rate of heat flow through a material. A high R-value indicates a low rate of heat transfer.

Salvage: The recovery of useful materials during renovation or demolition.

SBR: See Styrene-butadiene rubber

Scrim: A loosely woven fabric of natural or synthetic fiber, used in carpet and wall covering construction.

Secondary backing: Applies to tufted carpets only. The second part of the backing system, it is usually made of jute, and bonded to the primary backing with latex.

Secondary production: Refining, smelting, or processing raw materials into usable products. *See also* Primary production.

Secondary treatment: Wastewater treatment which generally digests nutrients and collects fine solids so that the water is safe to discharge. *See also* Primary treatment; Tertiary treatment.

Sedimentary rock: Formed from ancient silt deposits subjected to great pressure, cementing the deposits. Sandstone, limestone, and travertine are sedimentary.

Selvage: The finished edge on either side of a woven fabric, running the length of the piece. It is often of a different thread or weave, to prevent raveling.

Sheathing: Rigid sheet material used to cover the framework of buildings or cabinets.

Sheen: Reflectivity of a surface. Examples of sheen in finishes are:
Flat: A soft looking, velvety finish. Usually not washable.
Eggshell or satin: Slightly less sheen than semigloss.
Semigloss: Enough sheen to provide contrast to a flat finish and give greater resistance to wear and washing.
Gloss or high gloss: A very shiny and washable surface, more likely to show surface imperfections.

Shellac: Purified lac (a resin from a beetle), used for making varnishes and leather polishes. It is dissolved in methyl alcohol, and can also be thinned with safer ethyl alcohol.

Sick building syndrome: A pattern of health complaints related to poor indoor air quality. Symptoms include eye, nose, and throat irritation, nausea, fatigue, depression, headaches, and skin irritations. The symptoms disappear when away from the affected building for hours or days.

Silicates: Compounds based on the element silicon, a very common compound of sand and rock. Silicates are hazardous if inhaled but make very chemically stable building products.

Silicic acid (silica gel): A dehumidifying agent and rubber additive made from silica mineral. Low toxicity, but the dust is hazardous.

Siliconate: Any of a large family of chemicals based on silica, e.g. silicic acid, sodium silicate, silicone.

Silicone: Organic compounds of silicon used for caulkings and flexible plastics, lubricating oils and sealers. A very low toxicity material.

Silk-screening: A hand-print process using ink forced through a screen. A separate screen is required for each color.

Sizing: Essential before hanging wall covering on new plaster or gypsum, sizing is a liquid surface treatment that seals the surface against alkali, reduces absorption of the adhesive and provides "tooth" for the wall covering. Sizing also refers to a temporary, formaldehyde based, fabric treatment which makes fabric stiffer and easier to work with.

Slag: Waste from mineral or metal smelting operations. Slag is used in cement manufacture, lightweight concrete, and rock wool.

Slip-matched veneer: A method of matching veneer whereby the veneer is joined side by side to allow the pattern to repeat.

Sliver: A loose, soft, untwisted strand of textile fiber produced by a carding or combing machine, and ready to be spun into yarn.

Sludge: A clay-like residue containing very fine particles, resulting from wastewater treatment. Sludge is a disposal problem because it often contains toxic materials, e.g. lead, mercury and cadmium.

Slurry: A suspension, usually in water, of a finely ground solid, such as a metal or mineral ore.

Sodium fluorosilicate: A toxic compound used as a pesticide and preservative, and in dyeing processes. It is a popular mothproofing agent for wool.

Sodium silicate: Also called "water glass," a caustic liquid used for gluing cardboard, high temperature cements, and sealing asbestos fibers. Also used in foamed insulation.

Softwood: A wood from an evergreen, needle-bearing tree, such as pine, fir, hemlock, spruce, or cedar. Softwoods are fast growing and primarily used for construction.

Solution dyed: A dyeing method where the color is made an integral part of a synthetic fiber. The dyes are added to the fiber when it is still in a liquid form.

Solvents: Liquids capable of dissolving solids and maintaining them in solution. Most solvents used in paints, adhesives, and coatings are petroleum producers.

Soya lecithin: A fatty acid phosphate extract of soy bean oil, and also found in other grains and egg yolk. Used in inks, soaps, plastics, paints, and textile processing. Safe as a food additive.

Space dyed: A dyeing technique that consists of dipping parts of a long skein into different color baths to achieve a variegated filament.

Stack scrubbers: Devices which remove gases from industrial plant stacks. Sulfur dioxide scrubbers are the most common and help reduce acid rain emissions from coal-burning generators and smelters.

Staple yarn: Filament nylon that is cut into smaller lengths for easier handling. This yarn has a tendency to shed and pill, unlike BCF yarn, which is one continuous filament.

Stitch count: The number of rows lengthwise in a carpet, expressed as a quantity of stitches per inch. *See also* Gauge.

Stock dyed: The yarns are dyed before spinning.

Styrene-butadiene rubber (SBR): A synthetic latex formed from petroleum and used for carpet backings and elastic fabrics. SBR has a characteristic pungent odor and releases several irritating gases.

Subfloor: The structural floor under a finished floor. Subfloors are usually made from plywood, diagonal boards, or concrete.

Substrate: A material that provides the surface on which an adhesive is spread for any purpose, such as laminating or coating.

Sulfites and sulfates: Compounds formed with sulfur oxides. All are toxic, have very sharp or "rotten" odors, and cause acid rain.

Sulfuric acid: A potent acid used in chemical processing, storage batteries, dyes, and cellulose fiber manufacturing. Sulfuric acid is the largest volume industrial chemical made in the U.S. Its safe handling and disposal is a major industrial problem.

Surfactant: A surface-active agent, useful for its cleaning, wetting, or dispersing powers.

Sustainability: The practice of conservation and environmental protection which assures the availability of resources for future generations.

Synthetic: Produced by chemical methods, i.e. not extracted from natural sources.

Tack: The degree of stickiness, and consequently holding power, of an adhesive.

Tambour: Thin strips of wood laid parallel and usually backed with fabric, e.g. on a rolltop desk.

Task seating: Office seating designed to provide versatility in adjustments and comfort while working at a desk.

Tedlar: Tough, transparent fluoride plastic sheet used as a protective surface for wall coverings. It imparts resistance to yellowing, staining, corrosive chemicals, solvents, light, and oxygen.

Teratogenic: A substance which harms the developing fetus causing increased birth defects.

Terpenes: Organic, aromatic substances contained in the sap of softwoods. Terpenes are allergenic to sensitive people.

Terracotta: A hard-baked pottery used extensively in the decorative arts and as building material. It is usually made of red-brown clay, but may be colored with paint or baked glaze.

Terrazzo flooring: Marble or granite chips embedded in a binder that may be cementitious, noncementitious (epoxy, polyester, or resin), or a combination of both. Can be used with divider strips of brass, zinc, or plastic.

Tertiary treatment: Final wastewater treatment including destruction of active bacteria, rendering the water safe to drink. *See also* Primary treatment; Secondary treatment.

Textile wall covering: Fibers laminated to a backing material that provides stability and adhering qualities. The fibers can be natural or synthetic, or a mix, and woven or laid in parallel strands. Examples include grasscloth, strings, woven wool, linen, jute, and sisal.

Thermoplastics: Plastic resins which can be molded when heated and retain their shape when cooled.

Thermosetting: Plastic resins which become hard and rigid after hot molding, and cannot be softened again with heat.

Thickset method: Installation procedure used for uneven material, such as ungauged stone. The mortar bed is ¾" to 1¼" thick.

Thinset method: Installation procedure used for evenly gauged material, such as ceramic or granite tile. A dry set mortar as thin as ³⁄₃₂" is used, or an adhesive.

Thixotropic: A quality that allows a compound to soften under pressure, and harden when undisturbed.

Thread count: The number of warp ends and filling picks per inch in a woven cloth.

Tile: Stone cut into small, modular units for ease of transport and installation. The most common size is 1' × 1', but 1' × 2', or 2' × 2' may be available for some types of stones. Stone tile is generally around ⅜" thick.

Tile trims: Available in a wide variety of shapes, these are tiles formed with finished edges, to be used when the edge of a tile would be exposed. Trims include roundedges, coves, stair treads, bullnoses, etc.

Tip-sheared pile: A partial shearing of the looped yarns on the pile, creating a textured surface of cut and looped yarns.

Titanium dioxide: A white pigment used in paint, vitreous enamel, linoleum, rubber and plastics, printing ink, and paper. It has low toxicity but high covering power, brilliance, reflectivity, and resistance to light and fumes. Production creates large quantities of toxic waste.

Toluene: An aromatic component of petroleum with a strong solvent odor. Moderately toxic and used as a solvent for adhesives and inks.

Toxic: Any substance which causes harm to living organisms. There is a wide range of toxicity, from very low to extremely toxic.

Toxic waste: Waste material which is potentially harmful if discharged and which must be stored, treated, or neutralized for safety.

Traditional materials: Materials which have been used for several generations. Their properties and toxicity are known from experience. *See also* Generally regarded as safe.

Trichloroethane, 1,1,2 and 1,1,1: Potent solvents used in paints and inks, and as cleaning and degreasing fluids. 1,1,2 trichloroethane is sweet smelling and quite toxic. 1,1,1 trichloroethane, by contrast, is much less toxic.

Trichloroethylene: A solvent used in degreasing, dry cleaning, paints, and adhesives. Moderately toxic, and classed as hazardous waste.

Tufted: A carpet construction that consists of bundles of yarn pressed onto a mesh backing and then bonded, usually with latex. *See also* Primary backing. A secondary backing is needed for dimensional stability.

Tung oil: Oil obtained from the seed of the tung tree, widely used as a drying oil in paints and varnishes, and as a waterproofing agent. Tung oil has been associated with depression of the immune system.

Twill weave: Characterized by a diagonal pattern created by weaving the warp yarn under one and across 2 filling yarns. It is the strongest of the basic weaves.

Twist pile: A cut pile surface made from yarns tightly twisted and set by a special heating treatment, to increase wearability.

Ultraviolet (UV): Short wavelength, high energy, invisible light responsible for sunburn, skin cancer, and bleaching and deterioration of many materials.

Ultraviolet radiation protection: A substance added to finishes that helps the product resist discoloration from ultraviolet light.

Undercushion: A padding material laid prior to laying the carpet. It helps protect and prolong carpet life, and if not glued on, allows removal of the carpet for relocation and reuse.

Underlayment: A sheet material, usually wood or wood fiber, laid under resilient flooring, carpet, or tile to minimize irregularities in the subfloor or to add acoustic separation.

Underwriters Laboratories (UL): The U.S. testing agency responsible for verifying product electrical safety, fire ratings, etc.

Universal tinting system: A pigment coloring system used universally, achieving consistency of color from different sources.

Urea formaldehyde: An inexpensive polymer used extensively as glue for interior wood products. A source of toxic formaldehyde gas.

Urethane: See also Isocyanurate.

Value added: Secondary products with improved value over the raw materials. Value-added industries usually create jobs, and can make more efficient use of resources.

Vapor barrier: Material used to retard the passage of water vapor or moisture into insulated cavities.

Vat dyed: A dyeing method involving alkaline-soluble dyes, oxidized to produce color fastness in cellulosic fibers.

Veneer: A thin sheet of high grade wood formed by rotating a log and peeling a thin strip from it. The veneer is applied to thicker wood or paper to make plywood and decorative wood-surfaced panels for furniture and doors.

Vermiculite: A magnesium, iron, and silica mineral which expands when heated forming a lightweight, fire-resistant bead which is useful for insulation. The dust is hazardous to inhale.

Vinyl wall covering: A sheet of PVC vinyl, colored and usually embossed, and laminated to a backing that imparts stability and adhering qualities. The backing is usually paper, or a loosely woven film of polyester or cotton threads. Vinyl wall coverings are available in various face weights: Type I (12–19 oz. per lineal yard), Type II (20–28 oz. per lineal yard), and Type III (28–34 oz. per lineal yard).

Vitrified: A clay fired to the point of melting into glass-like substance (glassification).

Volatile organic compounds (VOCs): Gases with organic structures (based on the carbon atom), which are emitted from materials based on polymers or containing solvents, or plasticizers.

Volatile solvents: Solvents which evaporate at room temperature, adding to indoor air pollution.

Vulcanize: Transformation of a soft or liquid rubber into a very tough, heat resistant rubber usually by the application of heat and sometimes the addition of sulfur.

Waferboard: See Oriented strand board; Chipboard

Wallpaper: A printed paper wall covering with no backing or surface protective layer, generally for residential use.

Warp yarns: Backing yarns running the length of a woven fabric, parallel to the selvage.

Waste-to-fuel process: Burning of industrial waste to provide steam, heat, and often electricity to operate a plant. Wood waste and textile scraps are commonly burned for fuel.

Waterproofing: A treatment, such as a resin sealer or asphalt coating, that inhibits water penetration in building materials.

Water repellent: A surface treatment that imparts resistance to water penetration.

Weft yarns: See Filling yarn

Wilton: A woven cut or loop pile carpet using 100% wool or 80% wool and 20% nylon, with yarn through to the back. The backing is woven into the fabric of the carpet, eliminating the need for an adhesive to bond the face fabric to the backing.

Worsted (and semiworsted): A spinning process for long staple wool fiber, which produces a hard, twisted yarn.

Xylene: An aromatic component of petroleum with a sharp solvent odor. Xylene is moderately toxic and is used as a solvent for dyes, inks, paints, and adhesives.

Zinc oxide: A white pigment used in paints, ointments, plastics, and rubber which resists ultraviolet light and mold growth. Low toxicity, but the dust is hazardous.

FURTHER READING

BioLogic. David Wann. Johnson Books, Boulder, CO, 1990.

Building Construction Illustrated. Francis D.K. Ching. Van Nostrand Reinhold, New York, 1975.

Design for A Liveable Planet. Jon Naar. Harper and Row, New York, 1990. A review of global problems and recommended personal action.

Design for the Environment. Dorothy Mackenzie. Rizzoli, New York, 1991. A review of environmental issues as related to all disciplines of design.

A Dictionary of Textile Terms. Dan River Inc., New York, 1980.

Environmental Quality in Offices. Jacqueline Vischer. Van Nostrand Reinhold, New York, 1989. A guide to assessing office building performance from the occupants' point of view. Including case studies.

Fabrics for Interiors. Jack Lenor Larsen. Van Nostrand Reinhold, New York, 1975. A guide for architects, designers, and consumers.

A Glossary of House-Building and Site Development Terms. Canadian Mortgage and Housing Corporation, Ottawa, 1982.

The Good Wood Guide. Simon Counsell. Friends of the Earth, Ottawa, ON and Washington, DC, 1990.

Green Future. Lorraine Johnson. Penguin & Markham, New York, 1990. An account of the global issues and a detailed guide to personal action.

Healing Environments. Carol Venolia. Celestial Arts/Ten Speed Press, Berkley, 1988. A guide to the more spiritual aspects of homes and workplaces.

Healthful Houses. Clint Good and Debra Lynn Dadd. Guaranty Press, Bethesda, MD, 1988. A practical guide to building or renovating with safer materials.

The Healthy Home. Linda Mason Hunter. Rodale Press, Emmaus, PA, 1989.

The Healthy House. John Bower. Carol Communications, New York, 1989. A guide to the principles and practices of healthy house construction.

The Healthy House Catalog. Stuart Greenberg (ed.). Environmental Health Watch (4115 Bridge Ave., Cleveland, OH 44113), 1990. Articles on planning and building for health, including lists of products and services.

Interior Design. John Pile. Harry N. Abrams Inc., New York, 1988. An overall summary of the elements of interior design.

Interior Design and Decoration. 4th ed, Sherill Whiton. J.B. Lippincott Co., Philadelphia, PA, 1974.

Materials and Components of Interior Design. 2nd ed, J. Rosemary Riggs. Prentice-Hall Inc., New Jersey, 1989.

The Natural House Book. David Pearson. Simon & Shuster/Fireside, New York, 1989. An illustrated presentation of interiors designed with wood, stone, ceramics, glass, plant fibers, natural paints, etc.

NonToxic, Natural & Earthwise. Debra Lynn Dadd. Jeremy P. Tarcher, Inc., Los Angeles, 1990. How to protect yourself and your family from harmful products.

Paint Magic. Jocasta Innes. Frances Lincoln Publishing Ltd., London, England, 1981. A review of decorative paint techniques and materials.

Shopping for a Better Environment, Laurence Tasaday. Simon & Schuster, New York, 1992.

Shopping for a Better World, Rosalyn Will et al. Council on Economic Priorities, New York, 1989. A guide to manufacturers of consumer goods based on their corporate ethics in social, environmental, and political areas.

Tile Art, Noel Riley. Chartwell Books Inc., New Jersey, 1987. A history of decorative ceramic tiles.

Wood Users Guide, Pamela Wellner and Eugene Dickey. Rainforest Action Network, San Francisco, 1991.

Your Home, Your Health, and Well-Being, David Rousseau et al. Ten Speed Press, Berkley, and Hartley & Marks, Vancouver, BC, 1989. A practical guide to home construction and renovation for cleaner air, better lighting, and noise control, including special features for the environmentally ill.

ENVIRONMENTAL NEWSLETTERS & LISTINGS

The American Institute of Architects Environmental Resource Guide. 1735 New York Ave. NW, Washington, DC 20006.

Building With Nature. P.O. Box 369, Gualala, CA 95445. Professional networking newsletter.

Environmental Building News. R.R. 1, P.O. Box 161, Brattleboro, VT 05301. A newsletter on environmentally sustainable design and construction.

GREBE: Guide to Resource Efficient Building Elements. Steve Loken, Walter Spurling, and Carol Price. Center for Resourceful Building Technology (P.O. Box 3413, Missoula, MT 59806).

The Green House Bulletin. 4711 Springfield Ave, Philadelphia, PA, 19143. Monthly newsletter suited to the design and building community.

Interior Concerns Newsletter, Victoria Schomer. P.O. Box 2386, Mill Valley, CA 94941. A newsletter of environmental concerns for interior designers.

Interior Concerns Resource Guide. Victoria Schomer, P.O. Box 2386, Mill Valley, CA 94941. Product listings for healthier building materials.

Pacific Northwest Eco-Building Network Directory. Learned Integrated Habitats. (206) 850-7456. Regional resource guide of environmentally safe, energy efficient, nontoxic building materials.

Safe Home Digest. 24 East Ave., Suite 1300, New Canaan, CT 06840. A publication for healthy living.

Smart Wood Certification Program. Rainforest Alliance, New York.

INDEX

SUBSCRIPTION TO PROFESSIONAL EDITION

Available March 1, 1993

ENVIRONMENTAL *BY DESIGN* is also offered as a Resource Guide geared specifically to the building industry and design professional.

The PROFESSIONAL EDITION, in addition to the material in this book, consists of:

- **Semi-Annual Updates**
  In an ever-changing industry, current information is essential. The initial Product Reports will be supplemented with revisions and additions.

- **Manufacturer & Supplier Index**
  A comprehensive listing of the support network for all products listed, arranged by regions for the United States and Canada.

- **Information Bulletins**
  Diverse topics will be addressed in depth, such as reading Material Safety Data Sheets, pertinent environmental regulations and standards, and new developments and initiatives in related industries.

- **Material Summaries**
  Concise summary of properties for quick reference, to aid in selecting materials.

The PROFESSIONAL EDITION is in a looseleaf binder format, for easy inclusion of updates and bulletins. If you have purchased this volume and wish to supplement it with the additional material and services of the PROFESSIONAL EDITION, return it to us for upgrading.

When this copy is returned, $12US ($15CAN) will be discounted from the PROFESSIONAL EDITION, price of $40US ($50CAN). The subsequent yearly subscription price is $25US ($30CAN). Please enclose cheque for the discounted amount and send to Environmental *By Design*, P.O. Box 34493, Station D, Vancouver, BC, Canada V6J 4W4.

Name _____

Company (if applicable) _____

Address _____

_____

Profession	**Type of Work**
☐ Interior Designer	Residential ☐
☐ Architect	Offices / Banks ☐
☐ Contractor / Builder	Hospitality ☐
☐ Facility Manager	Institutional / Health Care ☐
☐ Dealer	Retail ☐
☐ Other _____	_____ Other ☐

REQUEST FOR INFORMATION
FOR PRODUCT REVIEW

ENVIRONMENTAL *BY DESIGN* is inviting requests from manufacturers and suppliers who would like their product reviewed for possible inclusion in the PROFESSIONAL EDITION and subsequent editions of this volume.

The information package includes:

- **Criteria and Standards**
   A description of criteria for the 14 symbols, and the minimum threshold for credit.

- **Questionnaire**
   A detailed survey of the product and manufacturer to aid in the review.

- **Checklist of Product Information**
   A list of standard product information to be submitted along with the questionnaire.

- **Full Outline of Product Categories and Materials Listing**
   A comprehensive list of interior finishing material categories in this publication.

Products successfully reviewed will be included in the PROFESSIONAL EDITION and in future editions of this volume. If you wish to have your products reviewed, return this form completed. Send to Environmental *By Design*, P.O. Box 34493, Station D, Vancouver, BC, Canada V6J 4W4.

THERE IS NO COST FOR A PRODUCT REVIEW OR LISTING

Company _____

_____

Address _____

_____

Name of contact person
best able to address environmental
issues for the product _____

Type of product(s) for review _____

_____

Requesting product review information does not automatically ensure that
the product will be included in future updates and editions.
Please enclose $10US ($12CAN) handling fee.

RELATED BOOKS
FROM HARTLEY & MARKS

YOUR HOME, YOUR HEALTH, AND WELL-BEING
by David Rousseau, W.J. Rea, MD, and Jean Enwright   300 pages
A practical, detailed guide to home construction and renovation for cleaner air, better lighting, and noise control, including many special features for the environmentally sensitive.

THE ENERGY CONSERVING HOUSE
A Guide to Super-Efficient Design and Construction
by Tom Lencheck, Chris Mattock, and John Raabe   224 pages
A revised and enlarged version of *Superinsulated Design & Construction*, covering every step in designing and building super-efficient homes, with detailed diagrams. An energy conserving house will save you money and reduce your domestic contribution to the greenhouse effect by 70%.

THE SHORT LOG & TIMBER BUILDING BOOK
A Handbook for Traditional & Modern Post & Beam Houses
by James Mitchell   240 pages
A valuable old/new building method using widely available and easy to handle short logs from smaller trees. Step-by-step diagrams for innovative designs, from a professional post and beam builder.

JAPANESE JOINERY
A Handbook for Joiners and Carpenters
by Yasuo Nakahara   240 pages
The only detailed, step-by-step guide to making traditional Japanese joints. Illustrated with hundreds of detailed diagrams, it reveals the once-secret guild methods used to make the joints.

MAKING TWIG FURNITURE & HOUSEHOLD THINGS
by Abby Ruoff   191 pages
Easy-to-make charming twig furniture from twigs and bark you can collect. 35 original designs, step-by-step, simple directions with diagrams. Includes guide to twig and bark sources.

SKILLS FOR SIMPLE LIVING
Edited by Betty Tillotson   224 pages
A collection of how-to letters from people who have always lived with environmental awareness, published over a period of twenty years in *The Smallholder*. Hundreds of practical ideas for reusing and transforming worn possessions, solving household and gardening problems, and much more, with numerous illustrations.

For a free catalog of all our books, write to:

Hartley & Marks, Inc., P.O. Box 147, Point Roberts, WA, USA 98281
                or
Hartley & Marks Ltd., 3661 West Broadway, Vancouver, BC, CANADA V6R 2B8